LIBRARY COPY
RACAL RESEARCH LTD

- 1 JUN 1998

WORTON DRIVE
WORTON GRANGE
READING
RG2 0SB, ENGLAND

SIMULATION TECHNIQUES AND SOLUTIONS FOR MIXED-SIGNAL COUPLING IN INTEGRATED CIRCUITS

THE KLUWER INTERNATIONAL SERIES IN ENGINEERING AND COMPUTER SCIENCE

VLSI, COMPUTER ARCHITECTURE AND DIGITAL SIGNAL PROCESSING
Consulting Editor
Jonathan Allen

Other books in the series:

MIXED-MODE SIMULATION AND ANALOG MULTILEVEL SIMULATION, Resve Saleh, Shyh-Jou, A. Richard Newton
ISBN: 0-7923-9473-9
CAD FRAMEWORKS: Principles and Architectures, Pieter van der Wolf
ISBN: 0-7923-9501-8
PIPELINED ADAPTIVE DIGITAL FILTERS, Naresh R. Shanbhag, Keshab K. Parhi
ISBN: 0-7923-9463-1
TIMED BOOLEAN FUNCTIONS: A Unified Formalism for Exact Timing Analysis, William K.C. Lam, Robert K. Brayton
ISBN: 0-7923-9454-2
AN ANALOG VLSI SYSTEM FOR STEREOSCIPIC VISION, Misha Mahowald
ISBN: 0-7923-944-5
ANALOG DEVICE-LEVEL LAYOUT AUTOMATION, John M. Cohn, David J. Garrod, Rob A. Rutenbar, L. Richard Carley
ISBN: 0-7923-9431-3
VLSI DESIGN METHODOLOGIES FOR DIGITAL SIGNAL PROCESSING ARCHITECTURES, Magdy A. Bayoumi
ISBN: 0-7923-9428-3
CIRCUIT SYNTHESIS WITH VHDL, Roland Airiau, Jean-Michel Berge, Vincent Olive
ISBN: 0-7923-9429-1
ASYMPTOTIC WAVEFORM EVALUATION, Eli Chiprout, Michel S. Nakhla
ISBN: 0-7923-9413-5
WAVE PIPELINING: THEORY AND CMOS IMPLEMENTATION,
C. Thomas Gray, Wentai Liu, Ralph K. Cavin, III
ISBN: 0-7923-9398-8
CONNECTIONIST SPEECH RECOGNITION: A Hybrid Appoach, H. Bourlard, N. Morgan
ISBN: 0-7923-9396-1
BiCMOS TECHNOLOGY AND APPLICATIONS, SECOND EDITION, A.R. Alvarez
ISBN: 0-7923-9384-8
TECHNOLOGY CAD-COMPUTER SIMULATION OF IC PROCESSES AND DEVICES,
R. Dutton, Z. Yu
ISBN: 0-7923-9379
VHDL '92, THE NEW FEATURES OF THE VHDL HARDWARE DESCRIPTION LANGUAGE, J. Bergé, A. Fonkoua, S. Maginot, J. Rouillard
ISBN: 0-7923-9356-2
APPLICATION DRIVEN SYNTHESIS, F. Catthoor, L. Svenson
ISBN:0-7923-9355-4
ALGORITHMS FOR SYNTHESIS AND TESTING OF ASYNCHRONOUS CIRCUITS,
L. Lavagno, A. Sangiovanni-Vincentelli
ISBN: 0-7923-9364-3
HOT-CARRIER RELIABILITY OF MOS VLSI CIRCUITS, Y. Leblebici, S. Kang
ISBN: 0-7923-9352-X

SIMULATION TECHNIQUES AND SOLUTIONS FOR MIXED-SIGNAL COUPLING IN INTEGRATED CIRCUITS

by

Nishath K. Verghese
Carnegie Mellon University

Timothy J. Schmerbeck
IBM. Rochester

David J. Allstot
Carnegie Mellon University

KLUWER ACADEMIC PUBLISHERS
Boston / Dordrecht / London

Distributors for North America:
Kluwer Academic Publishers
101 Philip Drive
Assinippi Park
Norwell, Massachusetts 02061 USA

Distributors for all other countries:
Kluwer Academic Publishers Group
Distribution Centre
Post Office Box 322
3300 AH Dordrecht, THE NETHERLANDS

Library of Congress Cataloging-in-Publication Data

A C.I.P. Catalogue record for this book is available
from the Library of Congress.

Copyright © 1995 by Kluwer Academic Publishers. Fourth Printing 1998.

All rights reserved. No part of this publication may be reproduced, stored in a retrieval system or transmitted in any form or by any means, mechanical, photocopying, recording, or otherwise, without the prior written permission of the publisher, Kluwer Academic Publishers, 101 Philip Drive, Assinippi Park, Norwell, Massachusetts 02061

Printed on acid-free paper.

Printed in the United States of America

For Kazuko - NKV

For Rosemary, Katie and Kim - TJS

For Vickie, Kevin and Emily - DJA

Contents

List of Figures xi

List of Tables xix

Preface xxi

1 Introduction 1

2 Sources of Noise and Methods of Coupling 5

  2.1 Semiconductor Device Noise and Phenomena 5
  2.2 Noise from Switching Voltage and Current 10
  2.3 Inductive Coupling 11
  2.4 Capacitive Coupling 30
  2.5 Substrate Coupling 34
  2.6 Summary 40

3 Semiconductor Device Simulation 43

  3.1 Significance 43

3.2	Basic Equations	44
3.3	Boundary Conditions	45
3.4	Models of Physical Parameters	47
3.5	Spatial Discretization	53
3.6	Solution Methods	62
3.7	A Representative Example	67
3.8	Summary	71

4 Simplified Substrate Modeling and Rapid Simulation ... 77

4.1	Simplified Equation	78
4.2	Spatial Discretization	80
4.3	Boundary Conditions	83
4.4	Solution Methods	84
4.5	Asymptotic Waveform Evaluation (AWE)	87
4.6	Substrate AWE Macromodels	92
4.7	Transient Simulation of AWE Macromodels	97
4.8	Substrate DC Macromodels	100
4.9	Matrix Solution	101
4.10	Results	107
4.11	Summary	114

5 Mesh Generation ... 117

5.1	Adaptive Mesh Refinement	117
5.2	A Priori Mesh Refinement	120
5.3	Summary	124

6 Substrate Modeling in Heavily-Doped Bulk Processes ... 125

6.1	Motivation	125
6.2	Single Node Substrate Model	127
6.3	Modified Single Node Substrate Model	129

6.4 Summary..133

7 Substrate Resistance Extraction for Large Circuits 135
  7.1 Nested Macromodeling 135
  7.2 Interpolated Macromodeling 141
  7.3 Summary..147

8 Modeling Chip/Package Power Distribution 149
  8.1 Effect of Power Bus Structure on Noise coupling....................149
  8.2 Summary..180

9 Controlling Substrate Coupling in Heavily-Doped Bulk Processes 183
  9.1 Characterization of noise coupling concepts............................184
  9.2 P+ Bulk Wafer Characterization...188
  9.3 Effect of Substrate contact placement on coupled noise 194
  9.4 Effect of Package Inductance on Substrate noise.....................196
  9.5 Noise Coupling Control Techniques...203
  9.6 Summary..214

10 Controlling Substrate Coupling in Bulk P- Wafers.. 217
  10.1 Bulk P- Wafer Characteristics 217
  10.2 Substrate Attenuation Structures 226
  10.3 Summary..231

11 Chip/Package Shielding and Good Circuit Design Practice 235
    11.1   Far Field Radiated Emissions 235
    11.2   Effect of Chip Signal Isolation/Shielding Techniques on Noise 240
    11.3   Effect of Packaging on Noise 244
    11.4   Effect of Card Layout and Referencing on Noise 248
    11.5   Effect of Circuit Topology on Noise......................................249
    11.6   Summary..252

12 A Design Example 255
    12.1   Design of a Mixed-Signal IC..255
    12.2   Summary..272

Appendices

A   Mesh Moments 275
B   Convergence Behaviour of Iterative Methods... 277

Index.. 279

List of Figures

Chapter 1

Chapter 2

2.1 Blow-up of a 5.5mm square chip wirebonded into a 68 pin PLCC package. 12
2.2 Side cross-section of a Plastic Dual In-line Package (PDIP) with an imbedded heat-sink. Courtesy of Amkor/Anam. 13
2.3 Cross section of MQFP/PQFP package. 14
2.4 Self Inductance of a wire loop in free space with round conductors..........15
2.5 Self Inductance of a round conductor over an ideal ground plane. 16
2.6 WIREBOND WIRE EXAMPLE -Self Inductance from bond pad to leadframe for 68 PLCC. (corner pin) 17
2.7 Flat, rectangular wire over a ground plane. 18
2.8 Routed signal and return wires. 19
2.9 Effect of Grounding the Package Leads. 21
2.10 The field fringing factor K_{L1}. 22
2.11 Mutual inductance of round or flat wires over a ground plane. 23
2.12 Effective loop inductance with opposing current directions 26
2.13 Effective inductance with opposing current directions; half of the loop..27
2.14 Supply decoupling capacitor designed to fit underneath a PLCC

	package.	28
2.15	Dimension drawing for 68 pin PLCC.	29
2.16	Microstrip fields	31
2.17	Scale drawing section of 0.8 micron process with interconnects.	33
2.18	Capacitive coupling (a) with and (b) without a close ground.	34
2.19	The general substrate coupling problem.	35
2.20	CMOS latchup.	37
2.21	Equivalent circuit of the latchup structure.	38
2.22	A typical Electrostatic Discharge protection circuit.	39

Chapter 3

3.1	Cross section of a MOS transistor.	45
3.2	Cell of a rectangular grid.	57
3.3	Local refinement in a rectangular and triangular grid.	58
3.4	Triangular grid and the boxes associated with them.	58
3.5	Gummel's algorithm.	63
3.6	Five-diagonal band matrix. Nx and Ny are the numbers of grid lines in the x-direction and y-direction respectively.	65
3.7	A representative example of the substrate coupling problem.	67
3.8	Impurity doping concentration versus depth in semiconductor. a) at the NMOS transistor gate and b) under the field oxide.	68
3.9	Impurity doping concentration versus depth in semiconductor. a) at the NMOS transistor drain and b) magnified view.	69
3.10	Transient simulation result showing the substrate coupling effect on the drain node of the sensitive transistor in Figure 3.7.	70

Chapter 4

4.1	A control volume in the box integration technique.	81
4.2	Resistances and capacitances around a node in substrate mesh.	82
4.3	Substate mesh connected to the electrical circuit.	84
4.4	Piecewise linear port voltage waveform.	97
4.5	Decomposition of the port voltage waveform.	98
4.6	Circuit schematic/layout profile for simulations with a heavily doped bulk and a lightly doped epitaxial layer.	108
4.7	Effect of various shielding techniques on peak-peak noise voltage at the	

LIST OF FIGURES

        sensitive node in Figure 4.6. 109
4.8   Effect of various shielding techniques on settling time of the noise voltage at the sensitive node in Figure 4.6. 109
4.9   Circuit schematic/layout profile for simulations with a lightly doped substrate. 110
4.10   Effect of various shielding techniques on peak-peak noise voltage at the sensitive node in Figure 4.9. 111
4.11   Effect of various shielding techniques on settling time of the noise voltage at the sensitive node in Figure 4.9. 111
4.12   Effect of various guarding configurations on the drain noise voltage of an NMOS transistor on the test chip [4.17]. 113
4.13   Effect of multiple substrate bias package pins /bond wires on (a) peak-peak noise voltage and (b) settling time of noise voltage. 113

Chapter 5

5.1   Interpolated (solid) vs. solution data (dashed) illustrating the importance of performing a solution between each level of refinement. 118
5.2   The peak-peak noise voltage at the sensitive node as a function of the number of grids in the z direction (with the grid density in the x and y directions fixed) for different substrate thicknesses. 120
5.3   The peak-peak noise voltage at the sensitive node as a function of the number of grids in the x and y directions (with the grid density in the z direction fixed) for different amounts of lateral separation. 121

Chapter 6

6.1   Circuit setup used to determine current flow in substrate. 126
6.2   Current flow lines in a heavily-doped substrate. 126
6.3   The single node model for a heavily-doped substrate. 128
6.4   Comparison of (a) peak-peak and (b) settling time behavior of the noise voltage as a function of the number of bonding pads used to bias the substrate in the experimental chip of Section 4.10.2. 129
6.5   Experimental setup to demonstrate limitation of single node substrate model. 130
6.6   Single node model for substrate of Figure 6.5. 130
6.7   Modified single node model incorporating lateral current flow. 131
6.8   Comparison of simulation results between the single node model and the

modified single node model. 131

Chapter 7

7.1 Unit cell model for a small block of substrate material.........................136
7.2 Representation of the "A" cell used to build the complete substrate model. 137
7.3 Four "A" cells and one "B" cell nested together. 137
7.4 Interconnection of "A", "B", and "C" cells. 138
7.5 Top view of the chip substrate showing "AB" cell locations..................139
7.6 A partitioning scheme for the substrate mesh. 141
7.7 Mesh boundary with four corners defined as external nodes/ports. 142
7.8 Lagrangian interpolation on the mesh boundary. 143
7.9 (a) Boundary currents and (b) Scale factors associated with each port for the boundary currents. 144
7.10 Example circuit divided into four partitions...145
7.11 Drain voltage of NMOS transistor in Figure 7.10.................................146
7.12 Partitioned simulation results vs. measured results. 146
7.13 Direct vs. Partitioned macromodeling..147

Chapter 8

8.1 RANDOM OR TREE (not recommended) 150
8.2 STAR (maximum isolation to pad) 150
8.3 Simple Grid (General Bus design) 151
8.4 Simple power GRID (for IC with wiring channels) 152
8.5 TREE TO GRID COMPARISON-bus electrical analysis. 153
8.6 STAR power feed using Grids. (Best of Star and Grid).........................154
8.7 Split power feeds using Stars. (equal to a split grid but a custom bus)....155
8.8 Communication between separate buses. (quiet to switching)................156
8.9 Driving logic to analog (noisy to quiet)-transient jitter.........................157
8.10 Driving logic to analog. (noisy to quiet) 157
8.11 Buffering Logic Signals Entering Analog Terrain. 158
8.12 Chip/Package power model for P+ bulk with P- epitaxial substrate. 160
8.13 Cross section of a Bicmos process with parasitic elements. 161
8.14 Chip example floorplan..163

LIST OF FIGURES

8.15 Example Chip Photomicrograph 164
8.16 68 PLCC Leadframe drawing with Power I/O Assignment. 165
8.17 Approximation of CMOS logic switching activity [8.14]. 166
8.18 Model showing individual pins and probing capacitance.[8.14]............167
8.19 Multiplied inverter model used to produce logic stimulus. 169
8.20 Power supply resonance when stimulated by a clock. 170
8.21 Power RLC ring noise versus I/O switching noise. 170
8.22 Connection of MOSCAP to get more capacitance to substrate from Gnd. 171
8.23 Cmos ground and Vdd ringing out of phase. 172
8.24 Substrate Noise As Logic Vdd and Gnd Capacitance is balanced. 173
8.25 Effect of Switching supply decoupling on Substrate noise. 174
8.26 Cmos power resonance coupled to substrate and analog power rails. ...175
8.27 Measurement configuration to measure supplies and SUBRING. 176
8.28 Substrate Coupled Frequency Spectrum with 1Mhz Clock. 178
8.29 Substrate Coupled Frequency Spectrum with 27Mhz Clock. 179
8.30 Output driver RLC tank circuit. 180

Chapter 9

9.1 Relative peak voltage noise. 186
9.2 Substrate Voltage Noise vs CMOS Switching Power. 186
9.3 Effect of Switching frequency on noise and power. 188
9.4 Necessity to thin wafer for effective high frequency backside connect. ...191
9.5 Single skin depth bounds where backside contact can be a ground plane. 192
9.6 Lateral resistance per square assuming chip edge contacts. 193
9.7 Substrate waveform without (top) and with(botttom) backside contact. .194
9.8 Substrate Contacts on Switching ground. (usually not recommended!) ..197
9.9 Substrate contacts on all ground busses; Forms resistive short! 197
9.10 Substrate Contact to Kelvin Top-Side Ground. (good!).........................198
9.11 Substrate Contacts on Non-Switching Ground. (good!).........................198
9.12 Substrate Contact on Kelvin (non power carrying) backside contact. ...199
9.13 Substrate Contact on power carrying backside contact. 199
9.14 Where to put substrate contacts for model of Figure 8.12. 200
9.15 Substrate RLC noise with substrate ties on all grounds. 201

xv

9.16   Substrate RLC noise with only analog substrate contacts. 202
9.17   Substrate output-coupled noise with analog sub contacts. 203
9.18   P+ substrate model for metallized ceramic pin grid array (PGA). 205
9.19   P+ substrate model for ceramic PGA with power planes. 206
9.20   Drawing of a metallized ceramic (MC) Pin-Grid-Array (PGA). 207
9.21   Drawing cross-section of metallized ceramic (MC) Pin-Grid-Array. 207
9.22   Drawing cross-section of multi-layer metallized ceramic PGA. 208
9.23   Blow-up of wiring from under flip-chip for MC/MCP PGA module. ...209
9.24   P+ substrate model for TAB package with dual ground plane..............210
9.25   TAB Package prior to being separated from the tape carrier. 211
9.26   P+ substrate model for C4 flip chip or chip on board connection.........212
9.27   Cross-section of a flip-chip on board packaging scheme. 212
9.28   P+ substrate model for 68PLCC with back side chip and card attache..213
9.29   C4 Ball difference between chip on FR4 verses chip on ceramic. 213

Chapter 10
10.1   Model for P- substrate. 218
10.2   Typical Bicmos implant profile. 221
10.3   Performance of different means to separate analog from digital on chip. 222
10.4   Top surface view of moats of Figure 10.3. 222
10.5   Simulated Moat Isolation Resistance with different structures. 223
10.6   Substrate Contact on all Power Carrying Grounds................................224
10.7   Substrate Contacts on Kelvin Ground. 225
10.8   Substrate Contact on Backside Contact...225
10.9   Simplified electrical equivalent model for substrate attenuator. 226
10.10 Substrate Splitting (voltage divider across substrate). 227
10.11 Substrate Splitting (less benefit with p+ buried layer). 228
10.12 Multiple substrate well isolation. 229
10.13 IDEAL SITUATION IS AN INSULATING SUBSTRATE. 230

Chapter 11
11.1   Radiated field from a magnetic dipole or current loop..........................237
11.2   Magnetic dipole radiated emissions envelope from square wave

LIST OF FIGURES

	source.	238
11.3	Common Mode Radiation from a package.	239
11.4	Electric Dipole Radiation Envelope.	239
11.5	Faraday shielding of circuit nodes.	240
11.6	Guard rings and well isolation in P- Bulk substrates.	241
11.7	Guard rings and well isolation in P+ Bulk/P- epitaxial substrates.	242
11.8	Nwell structure for carrier barrier, substrate contact, decoupling C.	243
11.9	Metal Quad Flat Pack or MQUAD Package. (Courtesy of OLIN).	245
11.10	Cross-section of a multi-layer, multi-plane, package.	246
11.11	Multi-chip module with 2 chips on common leadframe (8 PIN DIP).	247
11.12	Multi-chip module with 2 chip on common leadframe. (QFPK).	247
11.13	Difficulties Referencing chip signals off-chip.	249

Chapter 12

12.1	Product Design example architecture.	256
12.2	Product example overlap of successive waveforms.	256
12.3	Action of Timing and Gain Loops.	257
12.4	Product example block diagram.	258
12.5	Minimum noise corresponded with minimum switching activity.	259
12.6	D/A Converter Block Diagram.	261
12.7	Dac switch.	262
12.8	Problem ADC Topology.	264
12.9	Common mode to differential mode conversion.	265
12.10	Relationship of Synchronous Noise.	266
12.11	Net Affect of "Synchronous" Noise on ADC samples.	266
12.12	AGC output without (left) and with (right) logic switching.	267
12.13	The Value of "ON-CHIP" Decoupling.	268
12.14	P- Chip Floorplan with isolation regions shaded.	270
12.15	Simplified Equivalent Chip/Package model for Bus to Bus coupling.	271

List of Tables

Chapter 1

Chapter 2

2.1 Comparison of noise power bandwidth to 3db frequency of a N pole filter. 7
2.2 Inductive coupling from chip wires, bond wires, & 68 pin PLCC package leads. 30

Chapter 3

3.1 Mobility parameters for the analytical model. 48
3.2 Avalanche parameters. 52
3.3 Features of the 2 um BiCMOS technology 68

Chapter 4

4.1 Run-Time (on DECstation 5000) comparison between the device simulation program and the Macromodeling Techniques. 112

Chapter 5

Chapter 6

Chapter 7

7.1 Comparison of point-point simulated substrate resistance values with measurements. (All values in ohms, measured values + or - 10 percent) 140

Chapter 8

Chapter 9

9.1 CMOS logic noise with substrate ties on all grounds/analog ground..............204
9.2 CMOS I/O driver noise (Logic ground bus not tied to chip substrate). 204

Chapter 10

10.1 Implications of Scaling on Substrate Doping. 219

Chapter 11

11.1 Linear Dimensions vs Frequency in Air..236

Chapter 12

12.1 Resistance of island to island separation areas. 271

Preface

The goal of putting "systems on a chip" has been a difficult challenge that is only recently beginning to be met. Since the world is "analog" putting systems on a chip requires putting analog interfaces on the same chip as digital processing functions. Since some processing functions are accomplished more efficiently in analog circuitry, chips with a large amount of analog and digital circuitry are being designed. Whether a small amount or analog circuitry is combined with varying amounts of digital circuitry or the other way around, the problems encountered in marrying analog and digital circuitry are the same but with different scope. Some of the most prevalent problems are chip/package capacitive and inductive coupling, ringing on the RLC tuned circuits that form the chip/package power supply rails and off-chip drivers and receivers, coupling between circuits through the chip substrate bulk, and radiated emissions from the chip/package interconnects. To aggravate the problems of designers who have to deal with the complexity of mixed-signal coupling is the lack of verification techniques to simulate the problem. In addition to considering RLC models for the various chip/package/board level parasitics, mixed-signal circuit designers must also model coupling through the common substrate when simulating ICs to obtain an accurate estimate of coupled noise in their designs. Unfortunately, accurate simulation of substrate coupling has only recently begun to receive attention and techniques for the same are not widely known.

This book addresses two major issues of the mixed-signal coupling problem - how to simulate it and how to overcome it. It identifies some of the problems that will be encountered, gives examples of actual hardware experiences, offers simulation techniques and suggests possible solutions. Readers of this book should come away with a clear directive to simulate their design for interactions prior to building the design versus a "build it and see" mentality.

The first part of the book is a compendium of techniques to accurately model and simulate the substrate coupling problem, each with its pros and cons. Although it encompasses several ideas, the simulation techniques are for the most part results of several years of research at Carnegie Mellon University. Asymptotic Waveform Evaluation (AWE), a technique that will be discussed in lieu of simulating large linear circuits was developed at CMU by Ron Rohrer and his students, most notably Larry Pillage in the late '80s and early '90s. Nishath Verghese and David Allstot have been researching simulation techniques for substrate noise-limited performance in mixed-signal circuits at CMU since early '92 and are supported by the Semiconductor Research Corporation.

The second part of this book, is a result of struggles at IBM, Rochester, MN to build disk drive channel chips that operate between the few millivolt signals from the disk drive preamplifier and the microprocessors that accept the digitized read data and deliver the write data. Ideas are also included from problems encountered in serial optical links and phase lock loops operating on the same chips with digital processing functions. Tim Schmerbeck extends his thanks to the members of his team that helped him struggle through many of these problems. He also wishes to acknowledge the early pioneering work in mixed-signal coupling done at IBM by Larry Smith.

We are grateful to our respective spouses, significant others and families for their patience and understanding while we worked on this book. We also thank Hiok-Tiaq Ng for his review of the manuscript.

NKV

TJS

DJA

SIMULATION TECHNIQUES AND SOLUTIONS FOR MIXED-SIGNAL COUPLING IN INTEGRATED CIRCUITS

CHAPTER 1 *Introduction*

Recent product demands for smaller size have been met using multi-chip modules, chip-on-board (COB), and deeper integration ICs. These ICs are often required to combine significant portions of analog circuits with the purely digital switching functions. Current IC and packaging geometries have been miniaturized to the point that, at high speeds, even pure CMOS logic designs are being limited by crosstalk and inductive switching noise problems. Adding analog or any circuitry with less noise margin than CMOS logic to these ICs is a very difficult task. This is also the motivation causing many to put analog and digital on separate ICs. For some sensitive classes of circuits this may be the only alternative; separate ICs mounted in a multi-chip module, separate packages, or COB packaging. For a limited sub-class of analog circuits, the analog can be made to coexist with the switching functions. This may be at the cost of strict partitioning of switching and non-switching functions, extensive special handling, special semiconductor process, and a fully custom design effort. Since a single-chip solution is often the smallest, lowest cost and lowest power implementation, the additional effort is often justified. As the number of logic switching functions increases, the degree of difficulty merging the digital and analog functions will increase. Near field coupling between neighboring circuits, and coupling between widely separated circuits through the chip substrate and power rails are the big problems. As chip and package dimensions and clock frequencies increase, the wavelengths of the signals become comparable to chip and package interconnection lengths and this makes interconnections better antennas causing radiated emission problems from a single packaged IC. Radiated emissions problems can result from chip/package antenna lengths that are only 1/20th of the problem frequency wave-

Introduction

length or 1.5cm at 1Ghz. Exceeding FCC radiated emissions specifications usually occurs prior to manifestation of any actual functional concerns where the emissions perturb the design itself. Transmission line effects on silicon chips can't be neglected for 50ps rise times and wire lengths exceeding 3mm. R, L, and C couplings become more significant with higher packing density. Also, the scaling up of substrate doping with scaling laws makes physical partitioning of functions more difficult. As the total chip switching current increases it becomes very difficult to control inductive noise in the power lines. Card wires, package pins, bond wires, and IC interconnections all add inductance that cause fluctuation of the power lines and couple noise. The chip substrate acts as a collector, integrator and distributor of coupled noise on-chip. These chips usually require low inductance packages, shielding/isolation structures, on-chip decoupling, custom on-chip substrate ties, nwell ties, power, and signal routing with physical partitioning of function, custom chip floorplanning and every circuit noise rejection trick in the book [1.1].

On the verification front, crosstalk and coupling through interconnects has been the subject of much research. Most integrated circuit designers today use RLC models of chip, package and board level interconnects to more accurately analyze their designs for crosstalk and supply bounce. However, the simulation of noise coupling through the common substrate of mixed-signal systems has for the large part been ignored, mostly due to the difficulty in dealing with analysis of the substrate itself which is in effect a multidimensional interconnect connecting every transistor on the die with every other one. While mixed-signal designers have employed certain crude models for substrate coupling in the verification of their systems [1.2], accurate simulation of it has only recently begun to receive attention [1.3],[1.4].

We begin this book in Chapter 2 with an overview of the various sources of noise and the methods of noise coupling in integrated circuits (particularly mixed-signal ICs). Chapter 2 also serves as an introduction to the problem of substrate coupling. Chapters 3 - 7 focus on various techniques to both model the substrate and simulate the substrate coupling problem. In particular, Chapter 3 describes the semiconductor modelling and simulation strategies used in standard device level simulators. Chapter 4 describes a simplified model for the substrate and techniques to simulate the model in conjunction with a general circuit simulator. It outlines the Asymptotic Waveform Evaluation (AWE) technique, AWE based macromodels and circuit simulation using AWE macromodels. It also describes the use of dc (or resistive) substrate macromodels for faster simulation and compares simuation results using the macromodeling techniques with those obtained from a device simulation program and also with reported measurements on a test chip. Chapter 5 introduces the mesh generation techniques used in developing the substrate model. It introduces the ideas of automatic mesh generation and *a priori* discretization based on empirical results. In Chapter 6 we discuss a simple model that can be used for processes with a heavily-doped bulk and lightly-doped epitaxial layer. The frequently used single node model and its major limitation will be described. A simple extension to overcome the limitation of this model will also be described. Chapter 7 focusses on extraction techniques for large circuits. In the remainder of the book we will address solution techniques to overcome the various mixed-signal coupling problems. Chapter 8 will discusses modeling of power bus structures, its distribution and its impact on noise coupling. Experimental data will be used to describe the concepts and corroborate the models used. In Chapter 9 we describe techniques to minimize substrate coupling in heavily-doped bulk processes. Special attention will be given to the distribution of substrate contacts and to how the substrate is referenced in such processes. Chapter 10 describes solution techniques for lightly doped substrates. In particular, several shielding techniques and their characteristics will be described. In Chapter 11 we will talk about far field emissions and the issues involved in shielding the chip/package. The effect of packaging, shielding and circuit topology on noise will be discussed. Finally techniques for good circuit design will be recommended. Chapter 12 will give a run through the design process of a mixed-signal magnetic recording channel DSP chip and will illustrate the various issues discussed earlier.

REFERENCES

[1.1] Timothy Schmerbeck, "Mechanisms and Effects of Noise Coupling in Mixed Signal ICs", EPFL, Switzerland course presentation, June 29-July 10, 1992.

[1.2] T.J. Schmerbeck, R.A. Richetta, and L.D. Smith, "A 27 MHZ mixed A/D magnetic recording channel DSP using partial response signalling with maximum likelihood detection," *Technical Digest of the International Solid State Circuits Conference*, pp. 136-137, Feb. 1991.

[1.3] D.K. Su, M.J. Loinaz, S. Masui and B.A. Wooley, "Experimental Results and Modeling Techniques for Substrate Noise in Mixed-Signal Integrated Circuits," *IEEE Journal of Solid State Circuits*, vol. 28, no. 4, April 1993.

[1.4] N.K. Verghese, D.J. Allstot and S. Masui, "Rapid Simulation of Substrate Coupling Effects in Mixed-Mode ICs," *Proceedings of the Custom Integrated Circuits Conference*, pp. 18.3.1-18.3.4, May 1993.

CHAPTER 2 — Sources of Noise and Methods of Coupling

2.1 Semiconductor Device Noise and Phenomena

Clearly the lowest level noise present on semiconductor chips is due to electronic device noise caused by the random movement of charges through resistances, across transistor junctions, and random fluctuations in the charge recombinations in surface states and the semiconductor bulk. The level of noise generated and coupled by thermal noise, avalanche noise, shot noise, and 1/f noise represents a minimum level in coupled noise and all other noise mechanisms treated are usually orders of magnitude worse than these without special design. It is difficult enough to produce an amplifier with a rating of 1nano-volt/root Hz or less with device noise alone. Their control is accomplished mainly through optimum circuit design and topology with bandwidth limiting of signals and semiconductor process control. The circuit effects of chip thermal gradients, mechanical or piezoelectric stress, hot electrons effects, and mobile ionics such as sodium, can be considered very low frequency noise or noise coupling. Their control is usually accomplished with careful consideration of chip isotherms and mechanical stress lines, circuit design and biasing, balanced physical layout, and process control.

2.1.1 Thermal Noise (Johnson Noise)

The various particles within a resistor are constantly undergoing random thermal motion. This motion can be represented as lattice vibrations within the material. The

Sources of Noise and Methods of Coupling

lattice vibrations result in a random disruption of current flow. This can be thought of in terms of random charge carrier collisions with the lattice. The effective number of collisions is related to both the resistivity of the material and to the ambient temperature of the material.

$$En^2 = 4KTRB \qquad (2.1)$$

T = degrees Kelvin,

K = Boltzmann's constant,

R = resistance in ohms,

B = Frequency bandwidth in HZ.

$$En \approx \frac{\sqrt{RB}}{8} \mu volt \text{ at 25C} \qquad \text{R in M}\Omega, \quad \text{B in Hz} \qquad (2.2)$$

Expressed as an equivalent noise current:

$$In = \frac{Vn}{R} \approx \frac{\sqrt{B/R}}{8} \text{ picoamps at 25C.} \qquad \text{R in M}\Omega, \quad \text{B in Hz}. \qquad (2.3)$$

For example, a 1 mega-ohm resistor generates 1.25mvolt or 1.25 nano-amps over a 100 Mhz bandwidth.

The noise produced is WHITE NOISE or produces equal noise energy in any equal frequency interval. To first order the noise is current independent, i.e. not dependent on bias or signal current. There is a direct temperature dependence to this noise. As temperature goes up so does noise due to an increase in lattice vibrations. Reactive components do not inherently generate noise, only resistive ones. There is a direct resistance dependence. The increase in lattice collisions, which results in resistivity, causes an increase in noise level. The noise power bandwidth dependence is direct. Noise bandwidth is defined as the frequency span of a rectangularly shaped power gain curve that has an area equal to the power gain curve of the actual amplifier.

TABLE 2.1 Comparison of noise power bandwidth to 3db frequency of a N pole filter.

# Pole S(N)	3db roll-off freq. w/ N pole filter (Hz)	Fnoise/ F3db
1	6	1.57
2	12	1.22
3	18	1.15
4	24	1.13
5	30	1.11

Table 2.1 shows how much greater the noise power bandwidth of a system is than its 3db frequency depending on the sharpness of the roll-off at the 3db point or the number of poles in the roll-off.

2.1.2 Excess Thermal Noise

Excess thermal noise is that noise greater than is predicted by the theoretical expression for thermal noise. It occurs in resistor media due to disrupted current flow near grain boundaries. Generally it can be ignored for most resistor material formulations except for some carbon composition type resistors.

2.1.3 Shot Noise (Schottky Noise)

Current flow through a P-N junction is due to the movement of charge carriers through the depletion region. Microscopically, charge carriers traverse the P-N junctions as discrete elements rather than in continuous flow. The current looks like an average value with a superimposed ac component. The ac component is the shot noise. Bipolar base emitter current flowing through the base-emitter junction produces shot noise.

$$In^2 = 2qIB \quad \text{or} \quad In = 5.7 \times 10^{-4}\sqrt{IB} \quad \text{picoamps} \tag{2.4}$$

where:

Sources of Noise and Methods of Coupling

I = junction current in pa

B = bandwidth in HZ.

Shot Noise Example: A 1 microamp current (1 million picoamps) traversing a PN junction in a system with 100Mhz noise power bandwidth produces a noise current In= 5.7 nano-amps.

If the current is increased to 1milliamp with the same 100Mhz noise power bandwidth the noise current would increase to In= 0.18 micro-amps.

2.1.4 Flicker Noise (Pink Noise)

Flicker noise or 1/f noise as it is sometimes called decreases with increasing frequency and is usually a dominant device noise below 100HZ frequency and insignificant by a few KHZ.

The noise is generally associated with the surface effects of semiconductor materials and the generation and recombination of carriers that takes place at the surface. It is usually modelled as a current generator in the base of a npn, for example:

$$If^2 = \frac{K(IB^\gamma)}{F^\alpha} \qquad (2.5)$$

where:

I is the current in the base of the NPN transistor,

B is again the noise power bandwidth,

gamma is between 1 & 2 (usually 1,)

K varies between 1.2E-15 to 2.2E-12,

alpha is between 0.8 and 1.3 (usually 1) and

F is the frequency of the noise component.

2.1.5 Burst Noise (Popcorn Noise)

This noise component is usually only a few HZ to less than 1HZ in frequency. This low frequency noise has been related to heavy metal ions, especially gold, in the semiconductor and to imperfect processing. Some transistors jitter between two values of h_{fe} causing base current noise. It is the result of random recombination action in the semiconductor material or random jumping of carriers between two levels. It is usually a sign of poor processing. An empirical relation for bipolar transistors is shown below:

$$Inbst^2 = \frac{(KI^c B)}{(1 + (f/(fc))^2)} \qquad (2.6)$$

where:

I is the emitter current in an NPN transistor,

B is the noise power bandwidth of the system,

K is an empirical constan,

c= constant between 0.5 and 2,

fc = empirical burst noise cutoff frequency,

f= frequency of the noise component.

2.1.6 Avalanche Noise (improperly called zener noise)

The avalanche mechanism is that of high energy carriers creating additional electron hole pairs. The formation of the pairs is a discrete process and so forms discrete current spikes of random amplitude frequency and phase. The level of these current spikes tends to be geometry, and material dependent; and so a good general model does not exist.

A typical surface zener noise voltage at a current of 1ma is 100 nvolt/ ($\sqrt{Hz}$) (1mv for 100Mhz bandwidth). This is equivalent to the noise from a 600K resistor. Sub-surface or buried zeners are somewhat quieter (50 nvolt/ ($\sqrt{Hz}$)) and usually operate at somewhat lower current. Avalanche diodes have an essentially flat spectral response, however the amplitude distribution of the noise is not Gaussian. Devices that operate truly in the zener mode (tunneling rather than avalanche effect) exhibit shot noise only and are considerably less noisy.

2.1.7 Other Low Frequency Noise Mechanisms

Thermal feedback between circuits can be looked at as low frequency noise coupling with the frequency limited by the thermal time constant of the chip packaging system. This thermal time constant is seldom less than a hundred milliseconds. Piezoelectric effects and or mechanical stress effects on the integrated circuit chip also produce a low frequency noise effect. A chip packaged in an injection molded plastic package can have very large mechanical stresses placed on it. Equal stress lines form ellipses from chip edges that flatten toward the center of the chip. Stress can cause device mismatches if matching devices are on different equal-stress lines. ADC resistor ladder matching can be degraded by ~0.1% by placing the ladder near the chip edge. Package/card expansion/contraction can vary stress and therefore the piezoelectric effect giving rise to a form of low frequency noise.

2.2 Noise from Switching Voltage and Current

Clearly circuit voltage and current switching are the biggest noise sources on todays ICs and the simple, but often not practical, solution is to reduce the magnitude and frequency of current and voltage switching. The progression of logic design styles from best to worse in this respect are: (1) Steer current and sense current, eg., current mode analog (2) Balanced current steering with small voltage switches, eg., differential ECL (3) Unbalanced current steering with larger voltage switches, eg., unbalanced ECL (4) Switch current and switch voltage at less than supply swing, eg., TTL and (5) Switch maximum current and voltage while concentrating current and voltage transients during small time intervals and shorting out the supply during switching transients, eg., CMOS. CMOS, however popular, does represent near the worst choice as far as maximized current and voltage switching. For noise sensitive designs, viable alternatives to CMOS do exist. This is especially true at high frequency with high switching factor functional blocks that are the biggest noise producers and where the low static power of CMOS is not as important. Clearly techniques that reduce switching power such as shutting down all switching functions or logic drivers when not in use also reduce coupled noise.

2.2.1 Design styles from least to most noisy

- Steer current and sense current (current mode analog and digital.)
- Balanced current steering with small voltage switches (differential pair, differential ECL or CML or SCL (FSCL))
- Unbalanced current steering with larger voltage switches (unbalanced ECL or CML or CSL)
- Switch current and switch voltage (TTL)
- Switch maximum current and voltage (CMOS)
- Concentrate current and voltage transient during small time interval (CMOS)
- Short out the supply during switching transients, feedthru current (CMOS)

2.3 Inductive Coupling

Figure 2.1 shows a 5.5mm on a side chip wirebonded onto a leadframe in a 68 pin Plastic Leaded Chip Carrier (PLCC) package. The stamped metal leadframe contains the package leads as well as a metal die paddle or heat spreader for the chip to sit on. The chip backside is usually bonded with silver epoxy to the metal die paddle. The chip pads are wirebonded with gold or aluminum wire to the leadframe pins. A wirebond is usually added from the chip die paddle to a grounded leadframe pin to prevent coupling or radiation problems because of the floating die paddle metal. It is possible, when the leadframe is stamped out or etched, to combine the metal of several pins to form a lug pin. The lug pin has lower series resistance and inductance than the separate parallel pins since it includes in metal the separation area between the pins. For the same reason the lug pins are better thermal conductors than their parallel separate pin equivalent. The lug pin is shown directly connected to the die paddle as part of the leadframe design. The lug pins are typically connected to card ground. This provides a lower inductance electrical path and a better thermal conduction path. It is possible to connect lug pins on each side of the package to the chip die paddle. This "crab hand" connection to the die paddle provides a very low inductance path from the die paddle to card ground. Single package pins can also be connected to the chip die paddle in multiple places. Often a chip ground pad will be bonded out to a leadframe pin and the leadframe pin will be bonded to the die paddle. The low inductance of the die paddle provides a low impedance connection between widely separated chip ground pads.

Sources of Noise and Methods of Coupling

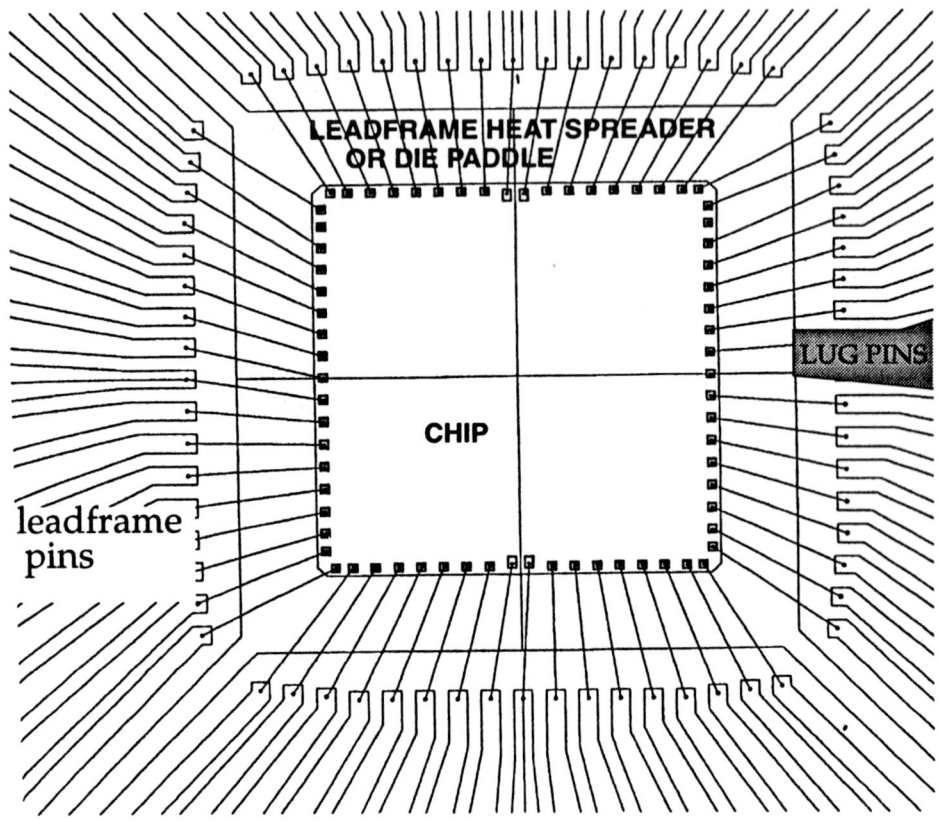

FIGURE 2.1 Blow-up of a 5.5mm square chip wirebonded into a 68 pin PLCC package.

Figure 2.2 shows a side cross section of a PDIP package. This particular package also imbeds a metal heat sink at the package bottom which is attached to the die attache pad or die paddle. This package metal heat sink can be a problem for coupling if it is not electrically connected in some way to a low impedance ac ground. This floating metal in the package provides a means for widely separated circuits on opposite sides of the chip and package to be coupled together via common capacitive coupling to the floating metal. The same situation holds for the doe attache metal of any common plastic package. Figure 2.3 shows a cutaway diagram of a Metric Quad Flat Pack

Inductive Coupling

(MQFP) or Plastic Quad Flat Pack (PQFP). The chip is sitting on the die paddle with the wirebonds shown connecting chip pads to leadframe pins. This picture also gives a feel for the scale of the wirebond and leadframe connection lengths in comparison to the chip dimensions. A long chip wire might be 1 millimeter; a wirebond wire could vary from 3millimeter to 6 millimeter (for standard 30.5micron diameter gold wire (1.2mils)) and the leadframe wire lengths could vary from 0.85cm (.33 inch) to 1.27 cm (.5 inch). Note that the chip pad spacings are roughly 0.15 millimeter (6 mils) while the leadframe spacings at the card end are 0.625 millimeters (24.6mils). The package serves the purpose of transferring the small spacings of the chip domain to the wider spacings allowed at the card level. The closer the lead spacings the higher the lead to lead mutual inductance. The highest mutual inductances per unit length, therefore, occur on the chip side of the package and get smaller as the lead fans out to the package edge. The bond wire have the smallest wire separations and the highest spacing above the card ground plane. Variations in bond wire to adjacent bond wire spacings do have a significant effect on the mutual inductances. All of the materials used in the package between and around leads are non-magnetic materials and therefore do not effect the inductances. The metal leadframe plating materials do however contain magnetic materials such as nickel but this has very little effect on the inductance.

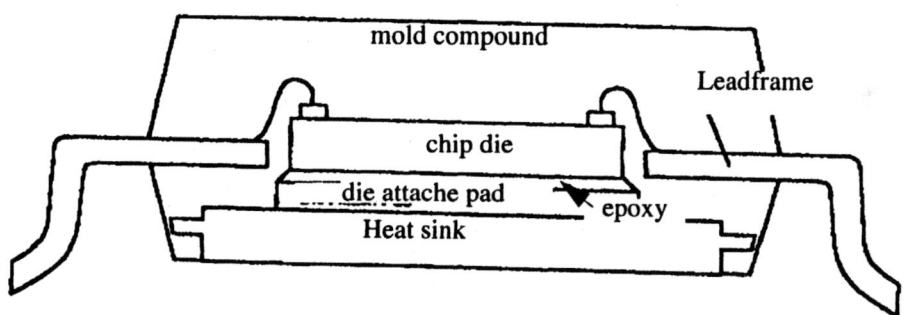

FIGURE 2.2 Side cross-section of a Plastic Dual In-line Package (PDIP) with an imbedded heat-sink. Courtesy of Amkor/Anam.

Sources of Noise and Methods of Coupling

Sources of Noise and Methods of Coupling

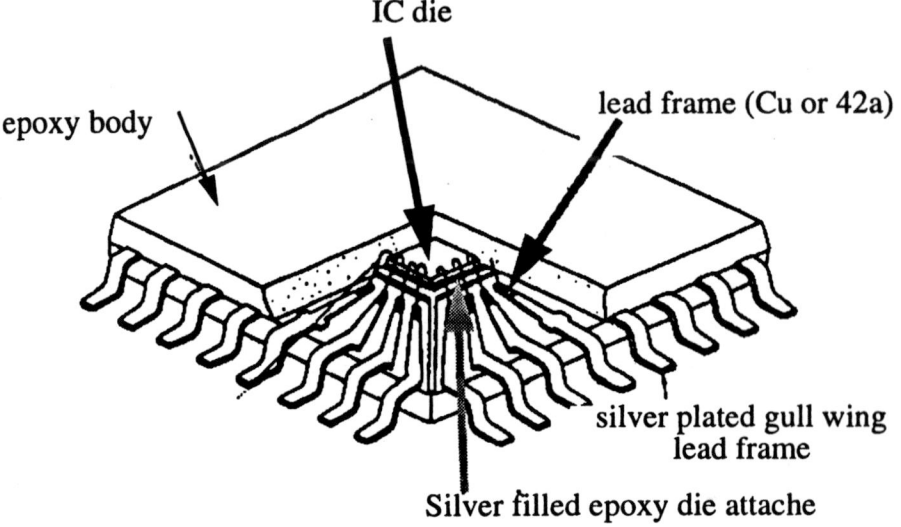

FIGURE 2.3 Cross section of MQFP/PQFP package.

In all of these packages the circuits on the chip are connected to the card or final carrier via very inductive connections. The absence of a ground plane in the package increases lead and wirebond spacing from ground and increases the package self inductance. The closer the leads and wirebonds are together and the farther they are spaced from a ground plane or grounded package lead the greater is the mutual lead to lead inductance. Plastic molded packages like these are among the most common and lowest cost IC packages. Lead pitch dimensions on the most aggressive of these packages have shrunk to about half a millimeter. An offsetting effect is that the height of the packages above the card ground plane has also shrunk to about 1 millimeter and the area of the package at a fixed lead count has also shrunk. This gives rise to lower self inductance but almost unchanged mutual inductance. The following sections will present some basic inductance relationships that will be applied to a 68 pin PLCC package to calculate its lumped self and mutual inductances. The equations presented

Inductive Coupling

will allow quick calculation of the inductance of most non-ground plane packages within 10 to 20% accuracy. The package self and mutual inductances give rise to one of the largest sources of chip noise coupling. [2.3]

2.3.1 Self inductive effects (In Free Space) for round conductors

Consider two circular conductors separated by distance d and of radius r: (conductors form a loop)

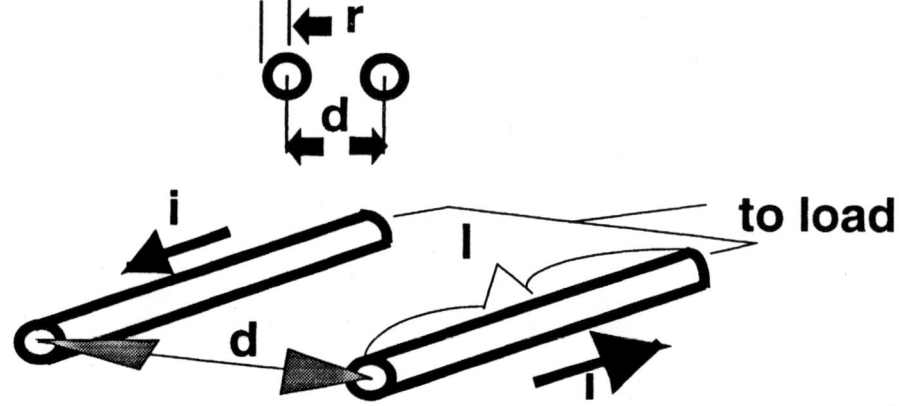

FIGURE 2.4 Self Inductance of a wire loop in free space with round conductors.

The external self inductance per unit length is approximated by

$$\frac{L_s}{l} = \frac{\mu_r \mu_o}{\pi} \ln\left(\frac{d}{r}\right) \quad \text{Henries/meter} \tag{2.7}$$

$$\frac{L_s}{l} \approx 0.4 \ln\left(\frac{d}{r}\right) \quad \text{nano-henries/mm} \quad (\text{for} \textit{nonmagnetic} \text{ materials}). \tag{2.8}$$

This assumes that most of the current flows at the surface of conductor and not much field inside the conductor. With this assumption it is not important whether the conductors themselves contain magnetic materials. This is only important for the regions between conductors that will carry the fields. The most common magnetic materials are iron, nickel, and cobalt which are not commonly used in IC packaging for other than plating or alloys within conductors. The assumption of nonmagnetic materials is

Sources of Noise and Methods of Coupling

Sources of Noise and Methods of Coupling

usually a good one. The self inductance is mainly dependent on the ratio d/r. The smaller the wire diameter the closer they must be together to keep the inductance per unit length from increasing. Put another way, the smaller the loop area the smaller the inductance.

2.3.2 Self inductance in proximity to a ground plane for round conductors

If a wire is over a ground plane, its inductance is approximately cut in half. The ideal ground plane return for the loop has theoretically zero inductance so one half of the loop path is effectively eliminated.

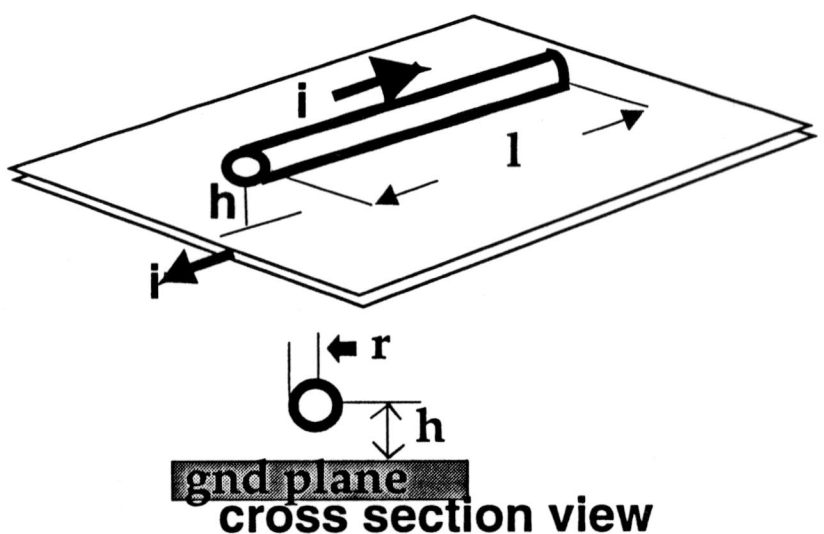

FIGURE 2.5 Self Inductance of a round conductor over an ideal ground plane.

$$\frac{L_s}{l} = \frac{\mu_r \mu_o}{2\pi} \ln\left(\frac{2h}{r}\right) \quad \text{Henries/meter} \tag{2.9}$$

$$\frac{L_s}{l} \approx 0.2 \ln\left(\frac{2h}{r}\right) \quad \text{nano-henries/mm (for} nonmagnetic\ materials\text{)} \tag{2.10}$$

Inductive Coupling

The self inductance is mainly dependent on the ratio h/r. The smaller the wire diameter the closer it must be to the ground plane to keep the inductance per unit length from increasing. Again, reducing loop area reduces the inductance.

Figure 2.6 uses equation (2.9) to calculate the self inductance of a wirebond wire from a 5.5mm square chip inside a 68 pin PLCC package that is assumed to be soldered on a circuit card with a ground plane. The shortest bond, a center lead, and the longest bond, a corner lead, are both calculated. As the chip size inside the package increases the difference between a center bond and a corner bond decreases. It is also assumed that none of the nearby module pins are tied to an ac ground so the closest ac ground potential is the card ground.

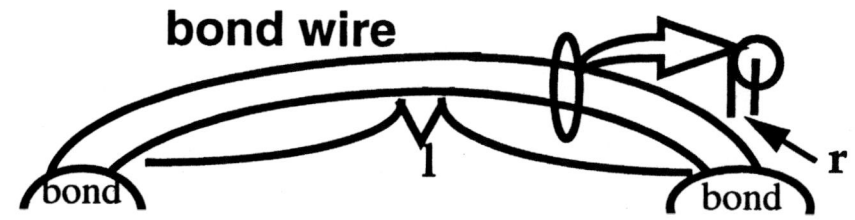

FIGURE 2.6 WIREBOND WIRE EXAMPLE -Self Inductance from bond pad to leadframe for 68 PLCC. (corner pin)

h = 0.1 inch =2.54 mm (average height with chip surface 0.9mm above leadframe)

r = 0.6 mils = 0.015 mm (radius of wirebond wire)

l = 0.1 inch = 2.54 mm   (length of bond wire)

(usually need to keep l less than 200 times the wire radius to prevent wire sag.)

$$L_s = 2.54 \times 0.2 \times ln\left(\frac{2 \times 2.54}{0.015}\right) \approx 3 \text{ nH}$$

For a middle pin we get:

$$L_s = 1.93 \times 0.2 \times ln\left(\frac{2 \times 2.54}{0.015}\right) \approx 2.24 \text{ nH}$$

The inductance per unit length is 1.165nh/mm. Note that since the wirebond wire has a curvature to it, the spacing between the wire bond and the ground plane is non-uniform. This will mean that the inductance will actually vary by a small amount with

Sources of Noise and Methods of Coupling

frequency. For frequencies below a hundred Mhz this can usually be neglected. The purpose of this exercise is to make the mathematics simple enough to get an insight into the physical attributes that affect the inductance values and to get a feel for the magnitudes.

2.3.3 Self inductance of flat, rectangular wire over a ground plane

In order to estimate the inductance of package leadframe wires or chip wires a flat wire over a ground plane needs to be considered.

$$L_s = \frac{\mu}{K_{L1}}\left(\frac{d}{w}\right) \text{Henries/Meter} \approx \frac{1.26}{K_{L1}}\left(\frac{d}{w}\right) \text{nH/mm}$$

d = spacing between wire and plane surfaces.

w = wire width

K_{L1} = fringing factor

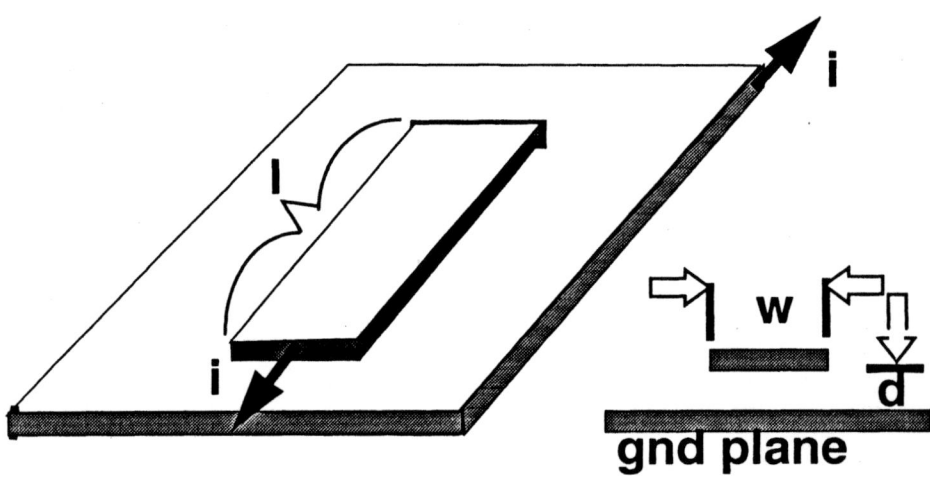

FIGURE 2.7 Flat, rectangular wire over a ground plane.

The fringing factor from Figure 2.10 is used to account for the fact that the fields are not all contained in the area between the flat conductor and the ground plane. In other

Inductive Coupling

words, it accounts for the amount of the field that is not contained between the two conductors.

- **EXAMPLE 1: CHIP WIRES**

d = 1micron. This assumes a chip ground plane within 1micron, or a low impedance wire spaced this distance above or below the signal wire. The silicon chip surface can act as a good ground plane if it is heavily doped. The wire width is assumed to be 3 microns; w = 3microns, $L_s = \frac{1.26}{1}\left(\frac{1}{3}\right)$ nH/mm or 0.42 nh for a 1mm chip wire length. The same inductance is obtained if a grounded signal return is routed on a metal level above or below the signal wire.

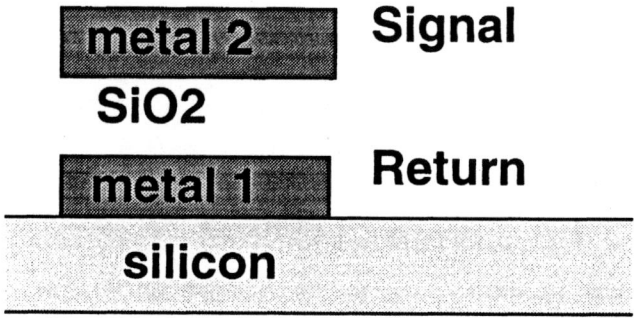

FIGURE 2.8 Routed signal and return wires.

The return wire routed above or below the signal wire can take the place of a nearby ground plane to reduce the inductance. This is usually the case for fairly isolated wires on a 2nd or 3rd metal interconnect layer. The grounded return wire in the above figure also reduces switching noise coupled to the substrate by shielding the chip substrate from the voltage transient. The transient couples to the return wire versus the chip substrate.

- **EXAMPLE 2: LEADFRAME WIRES (68 PLCC corner pin)**

Since the leadframe wires shown in Figure 2.1 actually change width an approximation of their average width is used. The leadframe length is broken up into a horizontal portion and a vertical portion which is at the edge of the package that actually touches the card. Again, we ignore the frequency dependence of the inductance, due to the non-parallel conductors, for simplicity.

d1=2.03mm (an average height above card for the horizontal leadframe portion)

Sources of Noise and Methods of Coupling

Sources of Noise and Methods of Coupling

d2=1.01mm (average height above card of vertical section)

w=0.7mm (average width value),

t=70um this is the average leadframe metal thickness (for information only),

L1 = 10.7mm (length of horizontal section) K_{L1}=3.8,

L2 = 2.03mm (length of vertical section), K_{L1}=2.4,

$$L_s = 10.7 \times \frac{1.26}{3.8}\left(\frac{2.03}{0.7}\right) + 1.01 x \frac{1.26}{2.4}\left(\frac{1.01}{0.7}\right) \approx 11 \text{ nH for a corner lead.}$$

For a center lead L1= 7.4mm $L_s \approx 7.9$ nH .

Total Self Inductance: (includes wire-bond inductance)

Corner Pin = 14 nh.

Center Pin = 10.1nh.

This calculation assumes that no nearby package pins are at ac ground. The length of the wirebonds and leadframe leads are strongly a function of the chip size and the design of the leadframe. A larger chip would result in less package inductance. However this would increase the potential for chip level inductance. Also inherent in the leadframe design is the spacing between the chip and the leadframe edge. Some leadframe designs result in nearly equal wirebond length because the spacing between the center of the chip and the leadframe is greater in the middle than at the corners. This leadframe style equalizes bond wire inductances but increases the difference between the center and corner lead inductances of the leadframe wire. The leadframe design considered here has equal spacing between the start of the leadframe and the edge of the chip. The chip considered here is also somewhat small for this package which yields higher package inductance.

Inductive Coupling

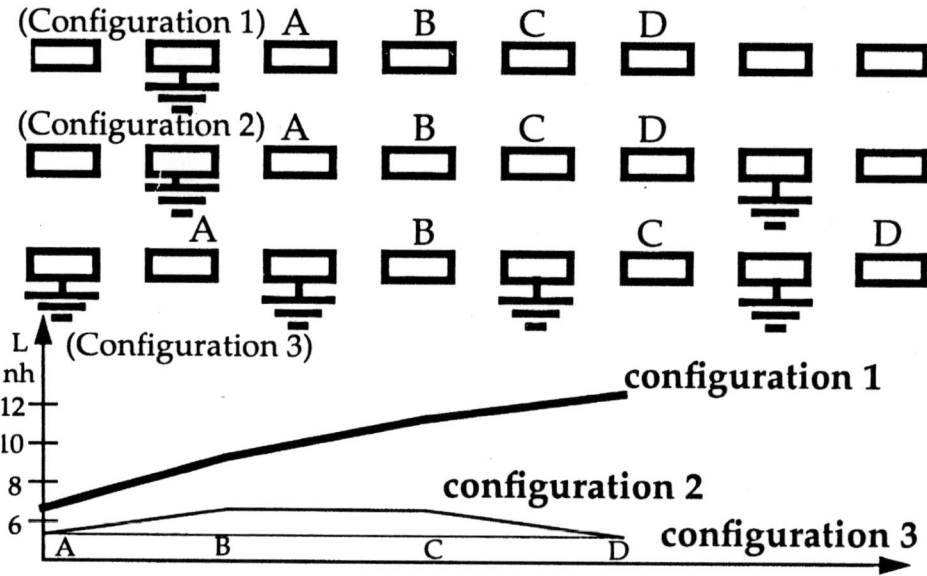

FIGURE 2.9 Effect of Grounding the Package Leads.

Figure 2.9 shows empirically measured total lead self inductance values for a 68 pin PLCC with various combinations of grounded leads. The leadsmeasured were in the center of the 17 leads on one side of the package. The leads are only grounded on the card side of the leadframe. Configuration 1 grounds a single lead and shows the inductance values for the four closest leads to the grounded lead. The inductance climbs rapidly as the spacing from the ground lead increases. Configuration 2 grounds a package lead on each side of the group of four signal leads. The inductance value and variation is reduced dramatically. Configuration 3 grounds every other lead in the package. The inductance value has been cut in half relative to the worst case in configuration 1 and the value is uniform. This large variation in possible package lead inductance is why the specified lead inductance of different manufacturers varies so wildly. It depends on the assumption of how many package leads provide a low impedance ac connection to ground. The impedance to ground of the various package leads will vary with frequency based on the circuits attached to them. This frequency

Sources of Noise and Methods of Coupling

dependent impedance to ground will introduce another frequency sensitivity to the actual self inductances.

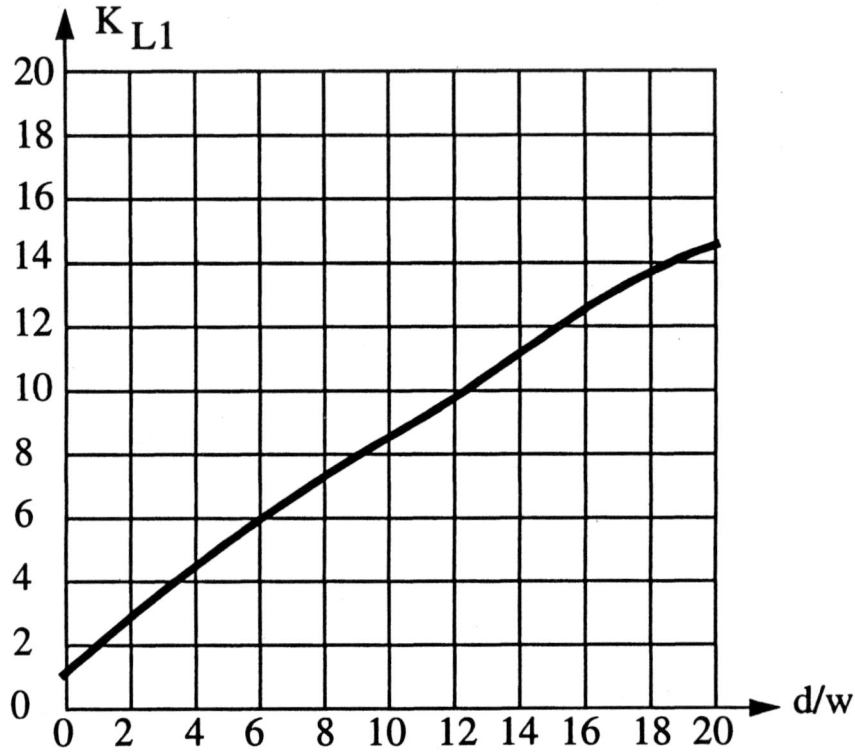

FIGURE 2.10 The field fringing factor K_{L1}.

2.3.4 Mutual inductance of round or flat wires over a ground plane

An approximate expression is shown in equation (2.11) below:

$$\frac{L_m}{l} = \frac{\mu}{4\pi} ln\left(1 + \left(\frac{2h}{d}\right)^2\right) \text{ Henries/meter} \qquad (2.11)$$

h = distance to gnd plane, d = distance between wires,

Inductive Coupling

or equivalently

$$\frac{L_m}{l} \approx 0.1 ln\left(1 + \left(\frac{2h}{d}\right)^2\right) \text{ nH/mm}. \tag{2.12}$$

FIGURE 2.11 Mutual inductance of round or flat wires over a ground plane.

Note that the mutual inductance is basically constant per unit length and is strongly dependent on the ratio h/d. The most effective means to reduce mutual inductance is to move the two conductors farther apart or closer to the nearest ac ground.

EXAMPLE 1: CHIP WIRES

h=1micron, d=3micron $\quad \frac{L_m}{l} = 0.1 ln\left(1 + \left(\frac{2 \times 1}{3}\right)^2\right) \approx 0.037$ nH/mm.

If we assume assume a length of 1mm then coupling on a second wire due to 10ma rms @ 50Mhz on the 1st wire is $V_2 = L_m\left(\frac{di}{dt}\right)$

$i_1 = 0.01 \sqrt{2} \sin wt$,

$\frac{di_1}{dt} = 0.01 \sqrt{2} w \cos wt$,

$|V_2| = L_m\left|\frac{di_1}{dt}\right| = 0.16$ mv peak.

This line to line inductive coupled value of 0.16mv is negligible in most applications and is usually reduced further by the presence of other wires in the neighborhood that have an ac impedance to ground. It is still significant when compared to device noise.

Sources of Noise and Methods of Coupling

WIREBOND WIRE EXAMPLE

h = 0.1 inch = 2.54 mm (average distance to ground plane),

d = 13mils = 0.33mm (average distance between wires in this example. Note that the chip pad center spacing may vary from 127micron on up and the leadframe pin spacing at the chip end is 0.45mm),

l = 0.1 inch = 2.54 mm (length of corner bond wire),

r = 0.6 mils = 0.015 mm (radius of wirebond wire),

$$\frac{L_m}{l} = 0.1 ln\left(1 + \left(\frac{2 \times 2.54}{0.33}\right)^2\right) \approx 0.55 \text{ nh/mm} = 1.4\text{nh for entire bond length.}$$

Coupling on second wire due to 10ma rms @ 50Mhz on 1st wire is

$$|V_2| = L_m \left|\frac{di_1}{dt}\right| = 6.2\text{mv peak.}$$

As expected the wirebond wire coupling is several times greater than the 1mm long chip wire. The wirebond wires have more than 14 times the mutual inductance per unit length than the chip wires. This is mainly due to the large increase in spacing from the nearest ground plane. Distances from a nearby ac ground on chip are never more than a few microns versus the tenth of an inch for the wirebond and leadframe wires. Since the spacing between wirebond wires is less than the spacing to the card ground, mutual inductance can be reduced significantly by the presence of nearby ac grounded wirebonds and leadframe pins. As we will see from the next example the wirebond wires also have 40% more mutual inductance per unit length than the package leadframe wires. This is primarily because the leadframe wires fan out and become gradually farther apart.

- Increase the current

The current of 10ma for mutual coupling purposes is probably not a realistic current value for a power supply lead current on a chip with large amounts of switching CMOS logic. If the current were 100ma instead we would get 62mv peak coupled to an adjacent wirebond wire. This is similar to the situation where a CMOS logic power wirebond is next to an analog bond wire. This doesn't even include the effect of the leadframe inductance. To get even this good you would have to directly wirebond the chip to the card. This is actually a common chip-on-board packaging scheme.

- Leadframe Mutual Inductance

Inductive Coupling

If the leadframe mutual inductance is taken into account:

d1 = 0.635 mm , d2=1.27mm (average lead separation for horizontal & vertical lead)

h1 = 2.03 mm , h2=1.01mm (average height of horizontal & vertical wire sections),

L1 = 10.7mm, L2=2.03mm (length of horizontal and vertical leadframe lead sections)

$$\frac{L_m}{l} = \frac{L_{m1}}{L1} + \frac{L_{m2}}{L2} = 0.1 ln\left(1 + \left(\frac{2 \times 2.03}{0.635}\right)^2\right) + 0.1 ln\left(1 + \left(\frac{2 \times 1.01}{1.27}\right)^2\right)$$

= .37nh/mm + .13nh/mm or 4.2 nh for entire lead length.

Coupling on second wire due to 10ma rms @ 50Mhz on 1st wire is:

$$|V_2| = L_m \left|\frac{di_1}{dt}\right| = 18.7 \text{ mv peak}$$

- Increase the current

For the case of a 100ma current in a CMOS supply lead the coupled noise due to the mutual inductance would be 187mv peak for the leadframe wire alone. Since on-chip decoupling lets more of the high frequency noise current stay on chip it lowers the amplitude and frequency of the noise current. This results in a reduction in the amount of coupled voltage. Low impedance power supply leads are not as affected by capacitive current coupling since it takes large currents to develop large voltages across low impedances. They are the most affected by series inductive voltage coupling which is independent of the net impedance. High impedance leads usually are most affected by capacitive current coupled to it from an adjacent voltage switching lead.

2.3.5 Interaction of self and mutual inductances

Note from Figure 2.12 and equation (2.13) that when the currents in the loop are in opposing directions like in a CMOS logic power supply loop that the mutual inductance actually reduces the effective loop inductance. For the previous example of a 68 PLCC package and two corner leads and wirebonds forming the loop with no nearby ac ground leads we would have an effective self inductance of:

$L_{eff-loop} = 2\langle 14 - 5.6\rangle = 16.8 nh$. This is a 40% reduction in loop inductance. By alternating additional ground and Vdd leads the effective loop inductance can be fur-

Sources of Noise and Methods of Coupling

Sources of Noise and Methods of Coupling

ther reduced. It is usually not desirable to alternate all the power and ground leads together at one position on the package. This increases the return loop area (and therefore the inductance) of the off chip driver load currents. A good compromise is to place the power leads in pairs at distributed positions on the chip and package. Note that switching noise coupled from noisy CMOS circuits to a separate set of on-chip power supply rails would be common mode to these rails. The mutual inductance would not reduce the self inductance for this common mode ac signal.

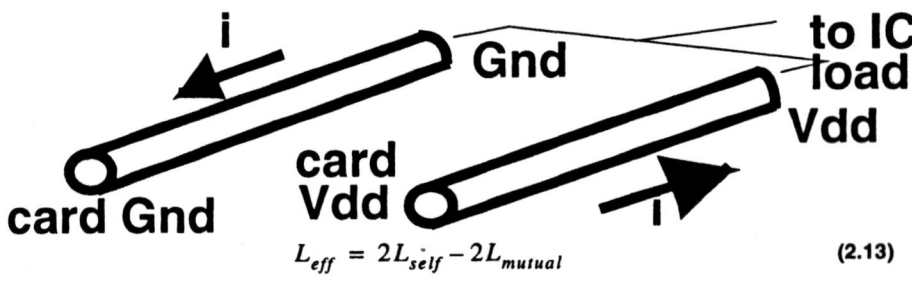

$$L_{eff} = 2L_{self} - 2L_{mutual} \qquad (2.13)$$

FIGURE 2.12 Effective loop inductance with opposing current directions

Note from Figure 2.13 and equation (2.13) that when the currents in the loop are in parallel directions the mutual inductance adds to the self inductance value. If package leads are paralleled to reduce inductance the mutual inductance dramatically limits the self inductance reduction. For the previous example of a 68 pin PLCC package and two parallel corner leads and wirebonds with no nearby ac ground leads we would

have an effective self inductance of $L_{self} = \frac{(14 + 5.6)}{2} = 9.8nh$. This is only about a 30% reduction in the self inductance. If the two parallel leads are separated by a large distance or on opposite sides of the package (like in the instance of power leads) the mutual inductance becomes very small so the total self inductance does cut in half. A loop formed of two wires in parallel with current in the same direction for each leg of the loop would have a loop inductance of 19.6nh. If only half of the pins are used but the currents are in opposite directions the loop inductance is 16.8nh. So 14% less inductance is obtained with half of the leads.

Inductive Coupling

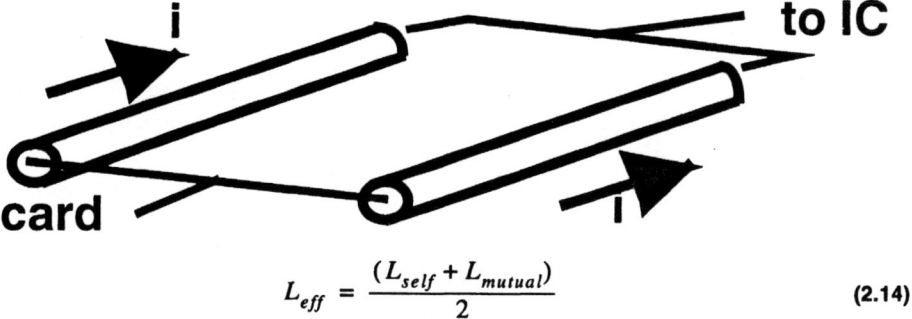

$$L_{eff} = \frac{(L_{self} + L_{mutual})}{2} \qquad (2.14)$$

FIGURE 2.13 Effective inductance with opposing current directions; half of the loop.

2.3.6 Package inductance effect on external card decoupling

If a 0.47µF ceramic capacitor on the card was connected between two leads of a PLCC the effective loop inductance would be greater than 16.8nh. The self resonance of this LC system at 16.8nh loop inductance is:

$$f = \frac{1}{2\pi\sqrt{LC}}, \quad f = \frac{1}{2\pi\sqrt{(16.8 \times 10^{-9})(0.47 \times 10^{-6})}} \approx 2Mhz$$

This means that even if the card capacitor was ideal its effective impedance would start increasing beyond 2Mhz. The rate of impedance increase would be 20db/decade. If the loop series resistance reached a minimum of 1 ohm at 2Mhz it would increase to 100 ohms by 200Mhz; which makes an off-chip capacitor not a very effective signal roll-off element at these frequencies. Since a good square-wave signal contains at least the first ten harmonics of the signal, a 20Mhz clock would carry signal harmonics up to 200Mhz. On-chip loop inductance can be obtained at a fraction of a nanohenry making them effective decoupling elements to extremely high frequencies. Bulk tantalum decoupling capacitors on card have even higher series inductances and resonate at even lower frequencies. Generally for a switching waveform, the on chip capacitances provide the current for decoupling during the rise and fall times, the card

Sources of Noise and Methods of Coupling

ceramic capacitors recharge the chip capacitors between edges over a few cycles and the card bulk capacitors recharge the ceramic capacitors over many cycles of the waveform. A failure in sizing the capacitors or in frequency response will result in an increased drop in the supply voltage due to switching transients. Figure 2.14 shows a supply decoupling capacitor that can fit under the PLCC package when it is soldered on a card. This capacitor structure can provide minimum external inductance between many package leads and the external capacitor.

FIGURE 2.14 Supply decoupling capacitor designed to fit underneath a PLCC package. Courtesy of Circuit Components Inc.

2.3.7 Example Summary

Table 2.2 summarizes calculated examples of self and mutual inductance of chip wires, wirebonds and 68 Pin PLCC leadframe wires together with examples of self inductive voltage bounce amplitudes and coupled mutual noise amplitudes. The mutual inductive coupling and self inductive bounce voltage amplitudes become quite large even at moderate currents. This is why switching and non switching functions

Inductive Coupling

can usually not share common or adjacent power or signal pins. Note that the self inductive bounce of a 1mm chip wire carrying 100ma @ 50Mhz rms is nearly 20mv. These levels of inductive bounce are normally not seen on chip due to the large numbers of capacitively coupled neighboring wires that effectively decouple the bounce voltage. This decoupling also results in the inductive bounce being distributed on the neighboring wires. The self inductive bounce of a package lead even with every other package lead grounded is still more than 200mv. Elimination of the small on chip coupling portion is possible with custom routed star power connections from switching and nonswitching circuits to the common connection. In many analog circuits the on-chip mutual inductance terms exceed device noise even for the 1mm chip wire. These coupled amplitudes from inductive effects are easily exceeded by capacitive current coupling for higher impedance circuit nodes. The mutual inductance, transformer coupled, voltages are most damaging on low impedance power supply nets that are unaffected by large capacitively coupled currents. For this reason, some on-chip capacitive decoupling is required to keep more of the supply transient current on-chip versus flowing through the bond and package inductances. Note that the total self inductance term drops from 11nh to 5nh by grounding every other lead on the package for the center leads. This has the effect of bringing the ground plane closer. Also note that the inductances calculated and measured were for a packaged soldered on a ground plane card. Putting a package in any kind of socket can dramatically increase the package spacing to the card ground plane and therefore dramatically increase the effective package inductance.

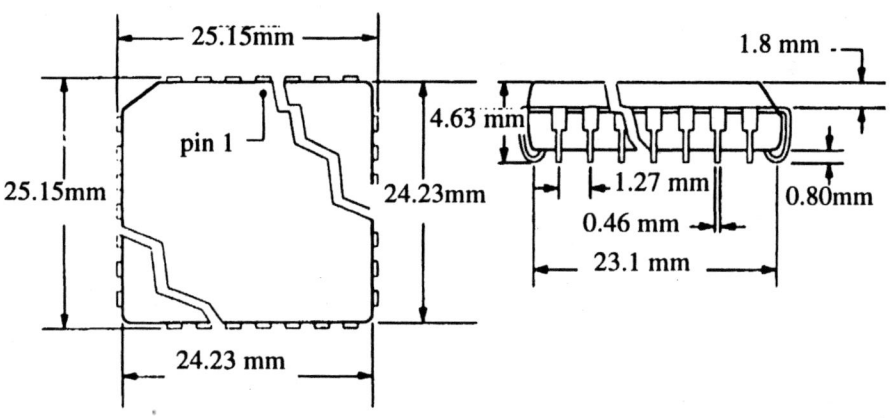

FIGURE 2.15 Dimension drawing for 68 pin PLCC.

Sources of Noise and Methods of Coupling

TABLE 2.2 Inductive coupling from chip wires, bond wires, & 68 pin PLCC package leads.

Conditions:	self inductance (nano henry (nh))	mutual-L (nano henry (nh))	mutual-L coupling from 10ma rms @ 50Mhz	mutual-L coupling from 100ma rms @ 50Mhz	self-L bounce from 100ma rms @ 50Mhz
1mm longchip wire. 3u width & spaced 1u above a ground plane	0.42	.037	0.16mv	1.6mv	18.6mv
Wirebond wire only, corner bond	3.0	1.4	6.2mv	62mv	133mv
Wirebond wire only, center bond	2.24				99.5mv
Leadframe wire only, corner lead	11.0	4.2	18.7mv	187mv	489mv
Leadframe wire only, center lead	7.9				351mv
Total corner lead & no nearby grounded pin	14	5.6	24.9mv	249mv	622mv
Total corner lead with every other package lead grounded	5.0				222mv

2.4 Capacitive Coupling

The inductance of a normal chip/package system is overwhelmingly dominated by the package. This is not always the case for the capacitance. The capacitance from a single lead on a 68 lead PLCC package to ground can vary from 2 to 4pf. The capacitance between adjacent leads is almost as large. Although this may be large compared to any single device capacitance on a chip, the capacitances of the composite chip between the power rails can be 100's to 1000's of pico-farads (pf). This capacitance is

Capacitive Coupling

composed mainly of oxide and junction capacitances. The capacitance of the metal interconnects becomes significant only for line to line coupling. However, 1000's of wires with say 0.05pf each to substrate can amount to a very large effective capacitance to substrate of tens to hundreds of picofarad (pf).

- MICRO-STRIP Vs. Chip Wires

Figure 2.16 shows a wire on a low loss insulator over a conductor. This microstrip configuration confines the E fields to the region between the conductor and ground plane. This is because of the lower dielectric constant of air. The propagation mode is close to ideal TEM. Chip wires terminate almost all of their field lines on neighboring wires. This causes cross talk and can excite multiple modes of propagation with slightly different transmission speeds giving rise to dispersion and degradation of the sharpness of edges as rise times drop. Figure 2.18 shows the capacitive coupling environment for wires immediately above the silicon substrate versus wires at higher levels[2.4].

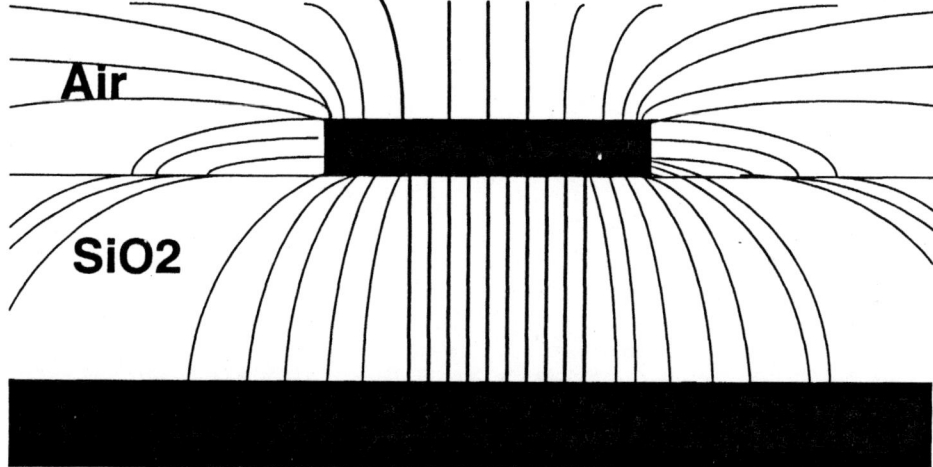

FIGURE 2.16 Microstrip fields

In Figure 2.18 part (a) the three signal wires are much wider than they are tall. The capacitance to the ground plane or chip substrate is much larger than the wire edge to edge capacitances. As a result most of the noise due to a transition on wire 2 is coupled to the ground potential and not to the other signal wires. The noise between wires is reduced at the expense of noise coupling to the chip substrate. For inverting logic families where there is on the average one rising edge for every falling edge, the sub-

Sources of Noise and Methods of Coupling

strate current injection is partially cancelled. For lower metal levels there is less line to line coupling but more coupling to and from the substrate. In Figure 2.18 part (b) the wires are either at higher interconnect levels above the substrate or above an insulating substrate. C12, C13, C14, and C15 are all of about the same magnitude. Due to the lack of a large ground capacitance in the denominator of the coupling equation the amount of coupled noise is increased. This multilevel connection system is worse than a single conductor over low impedance. This is similar to the noise coupling problems experienced on SOI (silicon on insulator) or SOS (silicon on sapphire) where the chip substrate does not act as a ground plane. As technology dimensions scale downward the wire width is scaled but the wire thickness is not scaled as much to keep resistance low. This increases wire to wire capacitance on a given level for adjacent wires. It doesn't make a lot of difference for non adjacent wires. Also, on modern process technologies, the level to level spacing is approaching the spacing on the same level. It is apparent that most chip wires will have many neighboring wires and many of these will look like ac grounds. These close proximity ac grounds will dramatically reduce the self inductance of the chip wires as discussed previously. For this reason, on chip self inductances are negligible for most IC applications until 100's of Mhz. This same environment that provides closely spaced wires for reduction of self inductance also provides for large capacitive coupling to occur. Sensitive wires must be isolated and/or shielded for capacitive switching transients.

For large chips with an inverting logic family like CMOS, there is, on the average, one rising signal for every falling signal. The total chip average of injected current into the substrate is much smaller than it would be for a non-inverting logic family. The major affect is that the substrate RLC circuit is stimulated. It is possible to get locally higher injection due to poor local cancellation. On smaller chips, a large area, single ended, locally interconnected, clock net routed on the first interconnect level above the substrate, can possibly inject large locally un-cancelled "output coupled" noise into the substrate. For this reason it is better to route on higher metal levels or use closely routed balanced out of phase clocks to reduce or cancel noise coupled to substrate. In a typical design of digital logic there will be periods of time when there is not good cancellation of current injected into the substrate and the phase of this coupling will depend on the logic switching activity, especially large off chip drivers.. Off chip drivers, pre-charge and self-resetting logic will tend to have much poorer averaging out of the coupling from output nodes to the chip substrate. Also, separation or shielding is necessary between noisy and sensitive wires. It is also necessary between the chip substrate or large p/n wells and noisy or sensitive wires. Since the separation between interconnect wire levels is approaching the separation distance between wires on the same interconnect plane, both vertical and horizontal shielding is required. Routing wires on adjacent vertical wiring levels at 90 degrees to one

Capacitive Coupling

another drastically reduces coupling between wiring levels. This is normally the case with digital wire routing but not as commonly with analog wire routing. Figure 2.17 shows a scale depiction of a four level metal, 0.8 micron photolithography, BiCMOS process. It is apparent that the metal land thicknesses are basically the same dimension as the metal to metal minimum spacings on a single metal layer. The spacings between metal on two adjacent levels are only 2 times greater than the spacing on the same level.

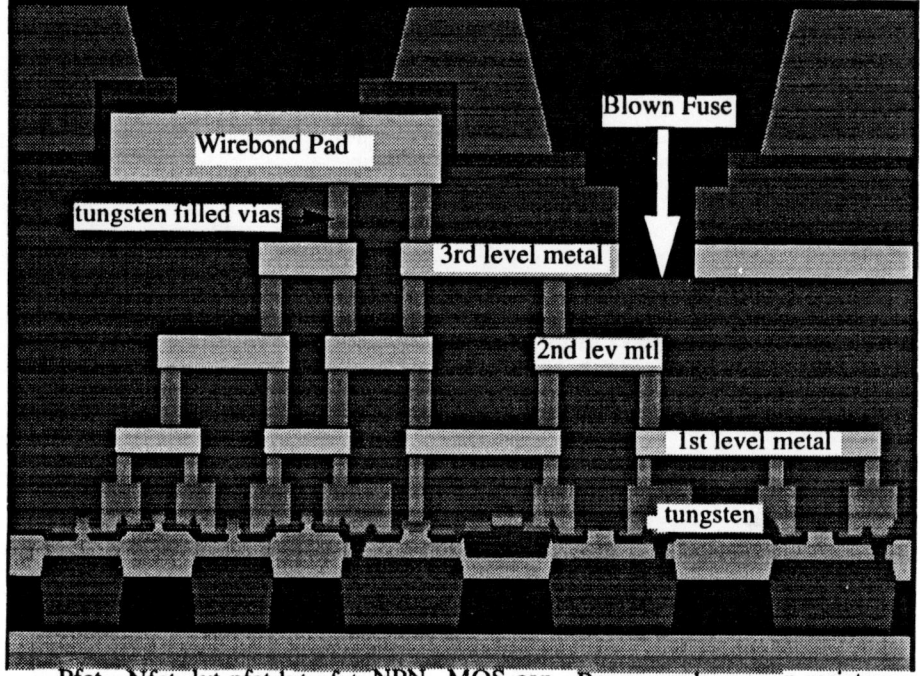

FIGURE 2.17 Scale drawing section of 0.8 micron process with interconnects.

An interconnect wire, polysilicon or diffusion tied to ac ground potential offers the maximum noise rejection for a given chip area if the shield can remain uncontaminated by the noise it is shielding from. If this can not be guaranteed, a physical separation of noisy and sensitive net is the best rejection solution. Reference [2.2] reports that in a 1.2μm BiCMOS process, leaving one empty wiring channel between two coupling wires provides 20 db of noise rejection.

Sources of Noise and Methods of Coupling 33

Sources of Noise and Methods of Coupling

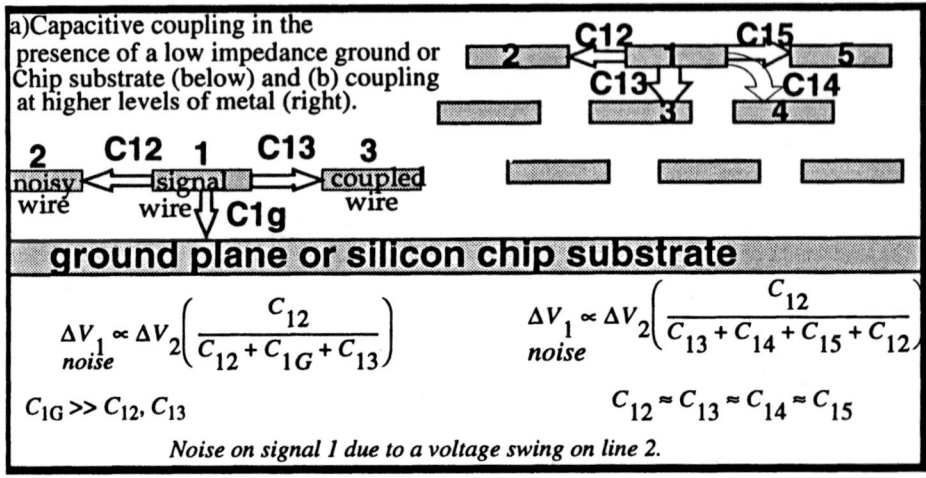

FIGURE 2.18 Capacitive coupling (a) with and (b) without a close ground.

2.5 Substrate Coupling

A major problem in the design of mixed-signal circuits today is that the digital and analog portions that are forced onto the same die actually interact with one another through the common substrate. With advances in circuit speed and new technologies such as the multichip module (MCM) technology the problem is likely to be aggravated.

2.5.1 Capacitive Coupling

The general substrate coupling problem is illustrated in Figure 2.19 in which digital switching nodes are capacitively connected to the substrate through junction capacitances and interconnect/bonding pad capacitances, causing fluctuations in the underlying voltage. As a result of this fluctuation, a substrate current pulse flows between the switching node and the surrounding substrate contacts. The switching induced impulse current flow causes the substrate potential underlying critical transistors in

Substrate Coupling

the path of this flow to change. One way in which these substrate potential fluctuations affect circuit performance is through the body effect.

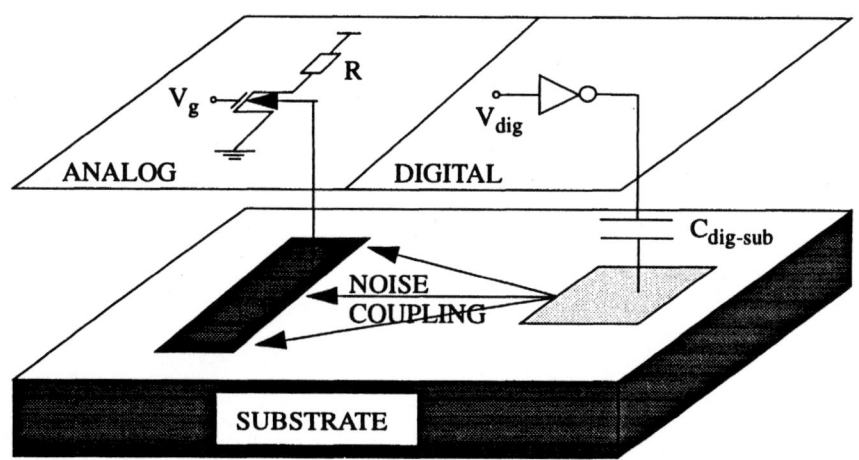

FIGURE 2.19 The general substrate coupling problem.

The threshold voltage of a critical transistor, V_t changes according to the formula (for NMOS transistors),

$$V_t = V_{t0} + \gamma\left(\sqrt{2\Phi_f + V_{SB}} - \sqrt{2\Phi_f}\right) \tag{2.15}$$

where V_{SB} is the source to body voltage of the transistor, V_{t0} is the zero bias threshold voltage, Φ_f is the Fermi level and γ is the body effect parameter. Accordingly, both the drain voltage and the drain current of the transistor vary so as to satisfy the transistor equations,

$$I_d = k_n'\left(\frac{W}{L}\right)\{(V_{gs} - V_t)V_{ds} - \frac{1}{2}V_{ds}^2\} \tag{2.16}$$

$$I_d = \frac{k_n'}{2}\left(\frac{W}{L}\right)(V_{gs} - V_t)^2(1 + \lambda V_{ds}) \tag{2.17}$$

Sources of Noise and Methods of Coupling

resulting in spikes of noise in their values. There is also a direct capacitive link between the substrate node and the drain, gate and source nodes of a sensitive transistor. Hence any fluctuations in the substrate potential translates into fluctuations in the drain current and could result in large spikes of noise at high impedance nodes.

2.5.2 Effect of Inductance

As current densities are increased and minimum feature sizes are reduced, the IR drops across the power distribution lines become more significant [2.5]. With larger amounts of current switching, the inductive noise associated with the power lines also increases [2.6]-[2.12]. This is especially important in high speed bipolar ECL chips and precharged CMOS circuits that require large current surges at the beginning of clock periods. These current transients can generate large potential drops due to the inductance of the power distribution network (EMF = -L dI/dt) This is referred to as power-supply-level fluctuations or as simultaneous switching, delta-I, ΔI ,dI/dt, and LdI/dt noise [2.4].

Until the advent of mixed-signal circuitry, power-supply-level fluctuations received more attention in high-speed, large I/O count bipolar circuits for mainframe computers because of the simultaneous switching of large numbers of fast output buffers in these systems. In mixed-signal circuits, however power-supply fluctuations are a significantly more critical determinant of system behaviour due to the presence of high impedance analog nodes that are extremely sensitive to noise in the power supply. Since the power supply (VDD or GND) is frequently contacted to the substrate to prevent latchup, any glitches in the power supply can couple through the common substrate directly onto the sensitive nodes. Unless precautions are taken to prevent or atleast minimize this coupling, power-supply fluctuations can be extremely detrimental to mixed-signal system performance.

2.5.3 Latchup considerations

Parasitic bipolar transistors are a problematic byproduct of all MOS processes. In CMOS processes these transistors are particularly troublesome because an n-p-n-p structure is formed by the n+ source of the NMOS transistor, the p substrate, the n well and the p+ diffusion of the PMOS transistor inside the n well (Figure 2.20). Due to the inherent positive feedback in this structure (shown in Figure 2.21), when it turns on, ground and power get effectively shorted together, large currents are produced and the circuit is destroyed. This is referred to as CMOS latchup [2.13]-[2.17]. The pnp transistor is formed by the p source of the PMOS transistor (emitter), n well (base), and p substrate (collector). The npn transistor is formed by the n well (collec-

Substrate Coupling

tor), p substrate (base), and n source of the NMOS transistor (emitter). R_{NWELL} and R_{PSUB} represent the n well and p substrate resistances to V_{DD} and GND respectively. When any of these two bipolar transistors is forward biased, it feeds the base of the other transistor, which in turn feeds the base of the first transistor, and this positive feedback increases the current until the circuit burns out. There are several ways of avoiding latchup and all of them focus either on reducing the gain of the bipolar transistor to weaken the positive feedback and on reducing the resistances R_{NWELL} and R_{PSUB} to prevent the parasitic transistors from turning on.

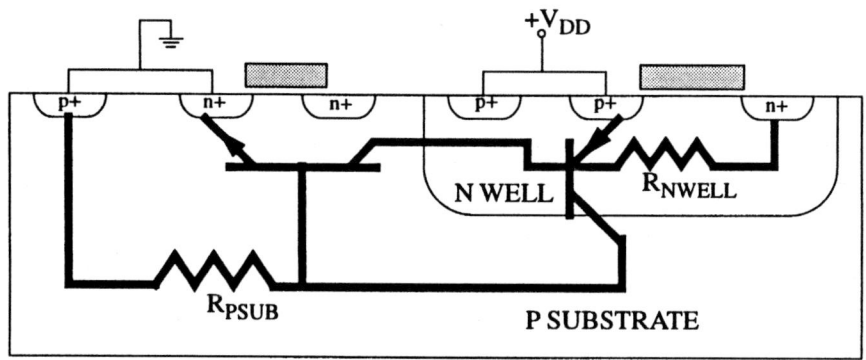

FIGURE 2.20 CMOS latchup.

Typical solution strategies to prevent latchup include placing guard rings around transistors (n+ connected to V_{DD} around PMOS transistors and p+ connected to GND around NMOS transistors) and contacting the wells and substrate periodically to either VDD or GND. The rationale behind these strategies is to minimize the parasitic resistances, R_{NWELL} and R_{PSUB}. Unfortunately, these solution strategies can prove to have deleterious side effects on noise coupling. Unless care is taken to prevent using the same power supply (ground) line for both switching circuitry and substrate connects and also in not using the same power supply (ground) lines for substrate connects in both switching and sensitive parts of the chip, noise could couple onto sensitive transistors placed far away from the switching sources, through the power supply. This could potentially aggravate the problem of noise coupling by creating a low resistance path (through the power supply) between the noisy and sensitive nodes in the circuit.

Sources of Noise and Methods of Coupling

Process-based technology improvements can also be employed to overcome the latchup problem and this involves the use of a lightly doped p epitaxial layer on a heavily doped bulk to reduce R_{PSUB}. Simultaneously, buried layers can be placed under n wells to reduce R_{NWELL}. This solution comes not only at added cost but also with the problem of across-chip substrate coupling. Since the bulk is heavily doped any noise injected into it by any single transistor spreads across the die influencing the behaviour of every other transistor. Simple solutions such as increasing the separation between sensitive and noisy nodes or placing local guard rings have little or no effect on noise coupling in such a process and consequently the problem is harder to deal with.

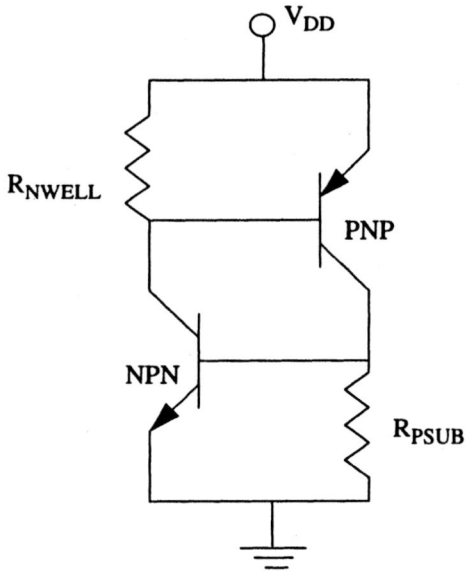

FIGURE 2.21 Equivalent circuit of the latchup structure.

2.5.4 ESD Considerations

Electrical overstress (EOS) and electrostatic discharge (ESD) are major reliability concerns in IC design and fabrication and are closely related to oxide breakdown and junction failures. Ungrounded conductive objects and people can accumulate static electric charges by induction or by contacting charged insulators. If these objects or people come into contact with an IC and are discharged through an ESD-sensitive path in the circuit, large momentary currents are generated and discharge through these paths [2.18]-[2.21]. The electrostatic discharge can damage the thin oxide layers and shallow junctions which are incapable of withstanding electric fields of high intensity.

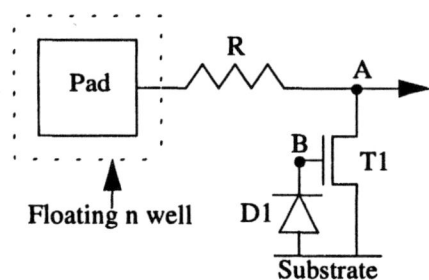

FIGURE 2.22 A typical Electrostatic Discharge protection circuit.

Inorder to protect an IC from ESD damage, input protection circuitry is used at all the pads that connect the IC to the external world. A typical ESD protection circuit is shown in Figure 2.22 where transistor T1 is the input protection device and diode D1 protects transistor T1 from electrical overstess [2.4]. During normal operation (within the power supply range) both transistor T1 and diode D1 are off. When the pad voltage surges out of the allowed range, some of the voltage is capacitively coupled onto node B preventing the gate oxide of transistor T1 from being stressed severely. As a consequence of this capacitive coupling diode D1 turns on and/or transistor T1 turns on and the input voltage is clamped. Since the discharging currents through transistor T1 and diode D1 flow directly into the substrate, n+ and p+ guard rings are typically placed around the protection circuitry to prevent latchup by picking up the injected carriers.

Besides latchup, ESD protection circuitry can also be a significant source of switching noise in mixed-signal ICs. Whenever the input voltages spike above or below the power supply range, the ESD protection circuitry is turned on and current spikes are injected directly into the substrate. Also, since the protection device (transistor T1 in Figure 2.22) tends typically to be large there is a significantly large capacitance from the drain/source nodes to the substrate, which can cause large currents to be injected into the substrate when the drain/source nodes switch. Even with the guard rings placed around them the noise injected into the substrate by the ESD protection circuitry can find its way to sensitive nodes in the IC, either directly through the substrate or through the power supplies that are contacted to the guard rings.

2.6 Summary

This chapter surveyed the various sources of noise including semiconductor device noise, as well as noise from switching circuit voltages and currents. Capacitive and inductive coupling of switched voltages and currents was considered. Consideration was also given to package inductance and capacitance with a 68PLCC taken as an example. Comparisons were given between coupling at the chip level vs the package level. Substrate coupling as a major source of mixed-signal coupling was introduced and the effects of inductance, latchup considerations and ESD protection devices on noise coupling was discussed. In the next few chapters we will deal with the substrate coupling problem and discuss techniques to model and simulate the problem.

REFERENCES

[2.1] Timothy Schmerbeck, "Mechanisms and Effects of Noise Coupling in Mixed Signal ICs", EPFL, Switzerland course presentation, June 29-July 10, 1992.

[2.2] L. D. Smith, *et al.*, "A CMOS-Based Analog Standard Cell Product Family," *IEEE Journal of Solid-State Circuits*, Vol. 24, No. 2, pp. 370-379, April 1989.

[2.3] Charles S. Walker, Capacitance, Inductance and Crosstalk Analysis, Artech House, Boston, 1990..

[2.4] H.B. Bakoglu, *Circuits, Interconnections and Packaging for VLSI*, Addison Wesley Publishing Company, 1990.

[2.5] W.S. Song and L.A. Glasser, "Power Distribution techniques for VLSI circuits," *IEEE Journal of Solid State Circuits*, vol. SC-21, no. 1, pp. 150-156, Dec. 1984.

Summary

[2.6] E.E. Davidson, "Electrical design of a high speed computer packaging system, " *IBM Journal of Research and Development*, vo. 26, no. 3, pp. 349-361, May 1982.

[2.7] A.J. Blodgett, "Microelectronic Packaging," *Scientific American*, pp. 86-96, July 1983.

[2.8] C.W. Ho, D.A. Chance, C.H. Bajorek, and R.E. Acosta, "The thin film module as a high-performance semiconductor package, " *IBM Journal of Research and Development*, vol. 26, pp. 286-296, May 1982.

[2.9] N. Raver, "FET off-chip drivers and package disturbs," *Proceedings of the IEEE Custom Integrated Circuits Conference*, pp. 574-579, May 1984.

[2.10] G.A. Katopis, "Delta-I noise specification for a high-performance computing machine," *Proceedings of the IEEE*, vol. 75, no. 9, pp. 1045-1415, Sept. 1985.

[2.11] I. Catt, D. Walton, and M. Davidson, *Digital Hardware Design*, Macmillan, London, 1979.

[2.12] J.H. Hackenberg, "Signal integrity in the VAX 8600 system," *Digital Technical Journal*, no. 1, pp. 61-65, Aug. 1985.

[2.13] D.B. Estreich, "The physics and modeling of latch-up in CMOS integrated circuits and systems," Ph.D. dissertation, Stanford University, 1980.

[2.14] D.B. Estreich and R.W. Dutton, "Modeling latch-up in CMOS integrated circuits and systems," *IEEE Transactions on Computer-Aided Design of Integrated Circuits*, vol. CAD-1, no. 4, pp. 157-163, Oct. 1982.

[2.15] R.R. Troutman, "Recent Developments and future trends in latch-up prevention in scaled CMOS," *IEEE Transactions on Electron Devices*, vol. ED-30, p. 1564, 1983.

[2.16] R.R. Troutman, "Recent Developments in CMOS latch-up," *Technical Digest of the IEEE International Electron Devices Meeting*, p. 264, 1984.

[2.17] J.Y. Chen, "CMOS- The emerging VLSI technology," *IEEE Circuits and Devices Magazine*, pp. 16-31, March 1986.

[2.18] R.K. Pancholy, "The effects of VLSI scaling on EOS/ESD failure threshold," *1981 Electrical Overstress/Electrostatic Discharge (EOS/ESD) Symposium Proceedings*, pp. 85-89, 1981.

[2.19] T.V. Mulett, "On chip protection of high density NMOS devices," 1981 Electrical Overstress/Electrostatic Discharge (EOS/ESD) Symposium Proceedings, pp. 90-96, 1981.

[2.20] C. Duvvury, " A summary of most effective electrostatic discharge protection circuits for MOS memories and their observed failure modes," *1983 Electrical Overstress/Electrostatic Discharge (EOS/ESD) Symposium Proceedings*, pp. 181-184, 1983.

[2.21] C.M. Lin, L. Richardson, K. Chi, and R. Simcoe, " A CMOS VLSI ESD input protection device, DIFIDW," *1984 Electrical Overstress/ Electrostatic Discharge (EOS/ESD) Symposium Proceedings*, pp. 202-209, 1984.

CHAPTER 3 *Semiconductor Device Simulation*

One of the available methodologies to simulate substrate coupling is a semiconductor device simulator such as TMA MEDICI [3.2] or MEDUSA [3.4] which employs numerical techniques to analyze semiconductor device action. In this chapter we discuss an overview of such an approach, its significance and its attributes [3.1].

3.1 Significance

Numerical simulation of semiconductor device operation is an important part of the design and manufacture of integrated circuits. It functions as a link between the theory of semiconductor physics and the complicated problems of integrated circuit fabrication and design, providing insights into device operation unavailable from purely analytical or experimental analysis. A semiconductor device is characterized by a set of physical parameters, e.g., mobility, lifetime, as well as technological parameters like geometry, impurity profile and so on. Simulation of the device begins with modeling which is the process of deriving from this set of parameters a field of electrostatic potential and quasi Fermi potential for electrons and holes in space and time. These three quantities yield, in turn, the vector fields of the electric field strength and electric current density. Finally, integration of the first vector along a contour between respective contacts and the second over respective contact areas results in the two terminal characteristics of a device.

3.2 Basic Equations

The electrical behavior of electronic devices is governed by electromagnetic field equations, which have to be supplemented by constitutive relations. For silicon devices with linear dimensions of a few micrometers or less and for frequencies up to 10^{10} Hz, the quasi-static approximation of the field equations is valid. These equations are the continuity equation for the electric current density J

$$\nabla J = -\frac{\partial \rho}{\partial t} \tag{3.1}$$

and Poisson's equation for the electrostatic potential ψ,

$$\nabla^2 \psi = -\frac{\rho}{\varepsilon} \tag{3.2}$$

where ε is the dielectric permittivity. Space-charge density ρ is composed of two different mobile carriers (electrons n and holes p) and the net impurity concentration N

$$\rho = q(p - n + N) \tag{3.3}$$

where q is the (positive) electronic charge and the net doping contains all ionized donors N_D and acceptors N_A

$$N = N_D - N_A \tag{3.4}$$

From solid-state physics it is well known that for semiconductors the total charge transport consists of electron transport in the conductivity band and hole transport in the valence band

$$J = J_n + J_p. \tag{3.5}$$

The splitting requires a separate continuity equation for each type of carriers

$$\frac{\partial n}{\partial t} - \frac{1}{q} \nabla \cdot J_n = -R \tag{3.6}$$

$$\frac{\partial p}{\partial t} + \frac{1}{q} \nabla \cdot J_p = -R \tag{3.7}$$

Boundary Conditions

where R is the net recombination rate. Irreversible thermodynamics teaches [3.5] that under isothermal conditions current densities for electrons and holes are proportional to the gradients of the corresponding electrochemical potentials or quasi-Fermi levels.

$$J_n = -q\mu_n n \nabla \phi_n \qquad (3.8)$$

$$J_p = -q\mu_p \nabla \phi_p \qquad (3.9)$$

where μ denotes mobility. Alternaltely, J_n and J_p can be written as as functions of Ψ, n and p, consisting of drift and diffusion components:

$$J_n = -q\mu_n n \nabla \Psi + q D_n \nabla n \qquad (3.10)$$

$$J_p = -q\mu_p p \nabla \Psi - q D_p \nabla p \qquad (3.11)$$

where D_n and D_p reflect the electron and hole diffusivities.

3.3 Boundary Conditions

The preceding field equations require boundary conditions in order to completely specify the boundary value problem to be solved.

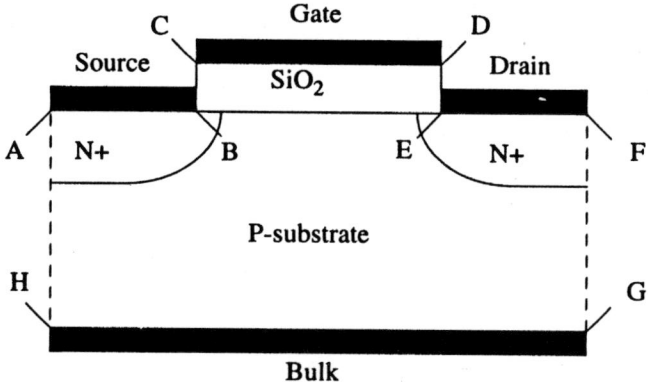

FIGURE 3.1 Cross section of a MOS transistor.

Semiconductor Device Simulation

To discuss the boundary conditions and its modeling consider the cross section of an arbitrary MOS transistor as shown in Figure 3.1.

Ohmic contacts (e.g., source, drain, and bulk) are often idealized by assuming infinite contact recombination velocities and space-charge neutrality. The contact is assumed to be a perfect source or sink for holes and electrons so that the semiconductor at the contact is in thermal equilibrium; hence, both the quasi-Fermi levels are equal to the voltage applied there

$$\phi_n = \phi_p = V_{appl}. \tag{3.12}$$

Likewise, we have Dirichlet boundary conditions for the electrostatic potential and for both carrier densities

$$\Psi = \Psi_0 + V_{appl} \tag{3.13}$$

$$n = n_0 \tag{3.14}$$

$$p = p_0 \tag{3.15}$$

where Ψ_0, n_0, and p_0 are the values of the corresponding variables for space-charge neutrality and at equilibrium.

For nonideal contacts such as Schottky contacts, the boundary condition for Ψ is still of the form (3.13), but Ψ_0 now includes a barrier height. Carrier concentrations at Schottky contacts depend on the current density across the metal semiconductor interface which is modeled as [3.6]

$$\mathbf{n} \cdot \mathbf{J}_n = -qv_n(n - n_0) \tag{3.16}$$

$$\mathbf{n} \cdot \mathbf{J}_p = qv_p(p - p_0) \tag{3.17}$$

where $\mathbf{n}$ is a unit vector normal to the surface. These equations reduce to (3.14) and (3.15) for infinite recombination velocities since n and p would stay fixed at n_0 and p_0.

For all Si-SiO$_2$ boundaries (e.g., line B-E in Figure 3.1) surface recombination is assumed where the normal components of both particle densities equal the surface recombination rate R_s

$$\mathbf{n} \cdot \mathbf{J}_n = -qR_s \tag{3.18}$$

$$n \cdot J_p = qR_s. \quad (3.19)$$

For the displacement vectors at both sides of such an interface we have the relation

$$n \cdot (\varepsilon_1 \nabla \Psi_1) - n \cdot (\varepsilon_2 \nabla \Psi_2) = Q_s \quad (3.20)$$

where Q_s is the sum of all effective net oxide charges per unit area at the Si-SiO$_2$ interface (oxide fixed and interface trapped charge). For bipolar devices or field oxide (3.20) can be simplified assuming zero electric field in the oxide layer, whereas for gate oxide two different approaches exist. One can either determine the field within the gate oxide region by solving Laplace's equation there, taking a Dirichlet boundary condition for the electrostatic potential at the gate contact into account; or one can assume a one-dimensional field perpendicular to the silicon surface.

At the remaining boundaries (e.g., lines B-C, D-E, F-G, and H-A in Figure 3.1) the normal components of electron and hole current density and electric field strength are assumed to be zero, yielding

$$n \cdot J_n = n \cdot J_p = n \cdot \nabla \Psi = 0 \quad (3.21)$$

Since this Neumann boundary condition is somewhat artificial, one has to decide from case to case if the induced error can be tolerated or not. However, this error can be made negligibly small by placing the corresponding boundaries at a distance far enough from the region of interest (e.g., lines F-G and H-A).

3.4 Models of Physical Parameters

The field equations contain quantities-e.g., the recombination rate R and the mobilities $\mu_{n,p}$ - which themselves are the result of rather complicated physical mechanisms. Therefore, these quantities are not constant but depend on the local values of current densities, and fields. Moreover, for the full set of equations, a relation between carrier densities, the corresponding quasi-Fermi potential, and the electrostatic potential is required. This relation results from carrier statistics and has to take into account band-gap narrowing and degeneration.

Though considerable differences exist for bipolar and MOS devices, the aforementioned physical mechanisms are generally valid and affect more or less most of the devices. Thus experimental results from special test structures can be evaluated yield-

Semiconductor Device Simulation

ing empirical relations, which describe these mechanisms and have been succesfully used in numerical simulations for a large number of devices.

3.4.1 Carrier mobility

Carrier mobility results from different scattering mechanisms. First of all, carriers are deflected by phonons and defects resulting in the relatively high bulk mobility of lowly doped material. This lattice mobility is reduced by additional Coulomb scattering at ionized impurity atoms and also by increases in temperature. The following empirical expression fitting experimental data for majority carrier mobilities in silicon [3.7],[3.8] and gallium arsenide [3.9],[3.10] and is both concentration dependent and temperature dependent has been given

$$\mu(N) = \mu_{min} + \frac{\mu_{max}\left(\frac{T}{300}\right)^\eta - \mu_{min}}{1 + \left(\frac{T}{300}\right)^\Psi \left(\frac{N_T}{N_{ref}}\right)^\alpha} \qquad (3.22)$$

where $N_T = N_D + N_A$ is the total doping concentration, T is the temperature and α, η, Ψ are constants. Usually, the same expressions are used for minority-carrier mobilities as well [3.1]. The typical values for the coefficients in (3.22) are summarized in Table 3.1 [3.2].

TABLE 3.1 Mobility parameters for the analytical model.

Parameter (Electron)	Silicon	GaAs	Parameter (Hole)	Silicon	GaAs
μ_{min}	55.24	0.0	μ_{min}	49.70	0.0
μ_{max}	1429.23	8500.0	μ_{max}	479.37	400.0
N_{ref}	1.072e17	1.69e17	N_{ref}	1.606e17	2.75e17
η	-2.3	-1.0	η	-2.2	-2.1
Ψ	-3.8	0.0	Ψ	-3.7	0.0
α	0.73	0.436	α	0.70	0.395

Measurements have shown that for for high excess carrier densities mobility is further reduced by electron-hole scattering [3.11],[3.12]. Following [3.13], the experimental data can be fitted replacing the total doping concentration in (3.22) by the expression

Models of Physical Parameters

$$N_T = 0.34 \cdot (N_A + N_D) + 0.66 \, (\bar{n}) \tag{3.23}$$

with $\bar{n} = p + n$ and using the coefficients of Table 3.1. This approach is justified by the fact that both mobility reductions are due to Coulomb scattering.

The fileld dependence measured in lowly doped material can be taken into account by the expression [3.7],[3.14]

$$\mu(N, \bar{n}, E) = \mu(N, \bar{n}) \cdot \left[1 + \left(\frac{\mu(N, \bar{n}) |E|}{v_{max}} \right)^{\beta} \right]^{-\frac{1}{\beta}} \tag{3.24}$$

with $\beta_n = 2$ and $\beta_p = 1$ resulting in doping-independent limiting drift velocities of

$$v_{max, n} = 1.1 \times 10^7 \frac{cm}{s}$$

$$v_{max, p} = 9.5 \times 10^6 \frac{cm}{s}.$$

Concering MOSFET simulations, channel mobility is further reduced by surface scattering or surface roughness. According to Yamaguchi [3.15] this effect can be taken into account by splitting the electric field strength into components which are longitudinal $(E_{||})$ and transverse $(E_\perp)$ to current density. In this case, $E_{||}$ is used in (3.24) and the resulting mobility is reduced by

$$\mu(N, E_{||}, E_\perp) = \mu(N, E_{||}) \, (1 + \alpha E_\perp)^{-1/2} \tag{3.25}$$

with $\alpha_n = 1.54 \times 10^{-5}$ cm/V and $\alpha_p = 5.35 \times 10^{-5}$ cm/V. The resulting maximum drift velocity, however, depends on $E_\perp$ and this is in conflict with reported measurements [3.16]. This problem can be avoided if the field independent mobility $\mu(N)$ is first reduced by $E_\perp$ using (3.25) and then the result $\mu(N, E_\perp)$ is substituted into (3.24) for drift velocity saturation. This however neglects the effect of carrier-carrier scattering as evidenced by the lack of the term $\bar{n}$ in (3.25). Other approaches for modeling carrier mobility are also possible [3.2].

3.4.2 Bandgap Narrowing and Degeneration

Fermi-Dirac statistics for both carriers yield the following relations between carrier concentrations, quasi-Fermi potentials and electrostatic potential:

$$n = \int_{-\infty}^{\infty} \frac{\rho_n dE}{1 + exp\left(\frac{(E-F_n)}{kT}\right)}, \qquad (3.26)$$

with $F_n = -q(\phi_n - \Psi)$

$$p = \int_{-\infty}^{\infty} \frac{\rho_p dE}{1 + exp\left(\frac{(E-F_p)}{kT}\right)}, \qquad (3.27)$$

with $F_p = -q(\Psi - \phi_p)$

where ρ_n and ρ_p denote the density of states functions. For nondegenerate semiconductors with a constant parabolic band structure these integral representations can be reduced to the classical Boltzmann approximation

$$n = n_i exp\left(\frac{\Psi - \phi_n}{V_T}\right) \qquad (3.28)$$

$$p = p_i exp\left(\frac{\phi_p - \Psi}{V_T}\right) \qquad (3.29)$$

where n_i is the intrinsic concentration.

In heavily-doped regions, the band structure of the crystal is no longer independent of the impurity concentration. The splitting of discrete energy levels of these impurity atoms into an impurity band, as well as the band-edge tailing narrows the effective bandgap E_G. This effect is of particular importance in bipolar devices because it strongly influences minority-carrier densities and currents within heavily-doped regions. Bandgap narrowing with a position-depending band structure and degeneration can both be taken into account in the integral expressions (3.26) and (3.27) without affecting the transport equations (3.8) and (3.9) [3.17],[3.18]. Typically bandgap narrowing is taken into account using the following expressions

$$n = n_{ie} exp\left(\frac{\Psi - \phi_n}{V_T}\right) \qquad (3.30)$$

Models of Physical Parameters

$$p = n_{ie} exp\left(\frac{\phi_p - \Psi}{V_T}\right) \qquad (3.31)$$

$$n_{ie}^2 = n_0 p_0 = n_i^2 exp\left(\frac{\Delta E}{kT}\right). \qquad (3.32)$$

Equations (3.30),(3.31) do not imply that Boltzmann statistics hold, rather (3.32) is used to model the electrical effects of of bandgap narrowing and degeneration simultaneously in an empirical fashion. The "effective bandgap narrowing" ΔE is taken from electrical measurements [3.19]-[3.21].

In [3.22] Selloni and Pantelides proposed a microscopic theory for bandgap reduction and identified sources of discrepancies among values of bandgap reductions that have been extracted from different experiments. In particular, they determined that the quantity ΔE appearing in (3.32) is given by

$$\Delta E = \Delta E_g + S(N, F, T) - D(N, T) \qquad (3.33)$$

where ΔE_g is the true bandgap reduction, S arises from statistics, and D arises from the disorder (fluctuations) in the impurity distribution. These two quantities tend to cancel each other, but their difference is not negligible (~20-30meV).

3.4.3 Recombination and Generation

For silicon, the net recombination rate R in (3.6),(3.7) is composed of Schockley-Read-Hall (SRH) recombination

$$R_{SRH} = \frac{np - n_{ie}^2}{\tau_p(n + n_t) + \tau_n(p + p_t)} \qquad (3.34)$$

Auger recombination

$$R_{Aug} = (c_n n + c_p p)\left(np - n_{ie}^2\right) \qquad (3.35)$$

and avalanche generation

$$G_{AV} = \alpha_n |J_n| + \alpha_p |J_p| \qquad (3.36)$$

In (3.34), n_t and p_t depend on the energy level of the traps and are given by the equations [3.2]

Semiconductor Device Simulation

Semiconductor Device Simulation

$$n_t = n_{ie}exp\left(\frac{ETRAP}{kT}\right) \quad (3.37)$$

$$p_t = n_{ie}exp\left(\frac{-ETRAP}{kT}\right) \quad (3.38)$$

where *ETRAP* represents the difference between the trap energy level E_t and the intrinsic Fermi energy E_i (i.e., $ETRAP = E_t - E_i$). Different experiments [3.23],[3.20] show that SRH lifetimes of electrons and holes decrease with increasing doping concentration. The empirical expression

$$\tau = \frac{(\tau_{max} - \tau_{min})}{1 + \left(\frac{N_T}{N_{REF}}\right)^\alpha} \quad (3.39)$$

with $0.3 < \alpha < 0.6$ is used successfully in numerical methods. A theoritical justification for an empirical relation like (3.39) is given in [3.24].

In contrast to SRH lifetimes, the Auger coeffiicients are nearly independent of doping, carrier densities, and temperature as indicated by different measurements [3.25],[3.26].

From Chynoweth's law it follows that the ionization coefficients α_n and α_p for avalanche generation depend exponentially on the electric field strength

$$\alpha = \alpha_\infty exp\left(\frac{-b}{|E|}\right). \quad (3.40)$$

TABLE 3.2 Avalanche parameters.

Electrons		Holes	
$\alpha_\infty \lfloor cm^{-1} \rfloor$	b (V/cm)	$\alpha_\infty (cm^{-1})$	b (V/cm)
7.03x10^5	1.231x10^6	1.582x10^6	2.036x10^6

Experimental data for the coefficients α_∞ and b [3.1] are shown in Table 3.2.

In addition to the recombination mechanisms described above, some simulators also include an additional recombination component at specific insulator-semiconductor interfaces [3.2]. This recombination mechanism can be described by a surface recombination velocity [3.27]. At the interfaces, an effective SRH lifetime for each carrier τ_n^{eff} and τ_p^{eff}, is computed based on recombination velocities v_n and v_p:

$$\frac{1}{\tau_n^{eff}} = \frac{v_n d}{A} + \frac{1}{\tau_n} \qquad (3.41)$$

$$\frac{1}{\tau_p^{eff}} = \frac{v_p d}{A} + \frac{1}{\tau_p} \qquad (3.42)$$

where τ_n and τ_p are the regular SRH lifetimes (possibly concentration dependent), A is the semiconductor area and d the length of the interface.

3.5 Spatial Discretization

In trying to achieve analytical solutions of the nonlinear basic equations, approximations are necessary, e.g., with respect to doping profiles, space-charge density, recombination models, etc. Though the resulting analytical models give some insight into physical device behaviour, the underlying assumptions often yield an oversimplified picture. This disadvantage can be avoided if the device equations are solved by numerical means.

The numerical solution of the governing partial differential equations comprises two fundamental steps, First, the space domain is mapped on a grid of distinct points or nodes. Applying some discretization method to the field problem then yields an algebraic problem with respect to space of finite though large size. Second, this problem is solved for the unknown variables. The discretized equations are obtained from the field equations by approximating the derivatives or integrals by expressions, which involve only the nodal values of the unknown functions. This can be accomplished by using interpolation functions in the neighbourhood of a node.

3.5.1 Finite Difference and Finite Element Methods

There is a variety of possibilities for deriving discrete equations for a field problem. Most common are either finite difference or finite element methods. Often these two methods are considered mutually exclusive from the very beginning. The finite difference method uses a local approximation of a differential operator and the finite element method applies a collection of shape functions as trial functions to approximate the desired solution globally, whereby residual (Galerkin) or variational (Ritz) strategies are used in the approximation process [3.28]. However for both methods, the starting point can be the field equations in differential or integral form or an equivalent variational problem. Likewise, the partitioning of the domain into subdomains and the use of trial functions is also applicable to both methods. It is generally accepted [3.28] that finite element methods always use trial functions which are defined subdomain by subdomain to have only compact support in the domain and which are then called shape or basis functions. This imposes a constraint on the choice of trial functions on the one hand and facilitates the adaption to the geometry of the domain on the other. In order to give some insight into the linkage between finite difference and finite element methods, they will be applied to Poisson's equation as a simple example.

Of the more common methods among the various finite difference techniques is the box integration method [3.29] which requires an integral representation of the governing equation, yielding

$$\int_v \nabla \cdot (\nabla \Psi) \, dv = -\frac{1}{\varepsilon} \int_v \rho \, dv \qquad (3.43)$$

which on applying the divergence theorem yields

$$\oint_{\delta v} \nabla \Psi \cdot dS = -\frac{1}{\varepsilon} \int_v \rho \, dv \qquad (3.44)$$

for any subdomain v. The subdomain results from a partition of the domain into "boxes" without overlap or exclusion, where each box includes one grid point. In the one-dimensional case, the box for a grid point x_i is the interval[$(x_{i-1} + x_i)/2$, $(x_i + x_{i+1})/2$], where x_{i-1} and x_{i+1} denote the neighbouring grid points.

The resulting difference equations are obtained from approximations of the box integral in (3.44). For the one-dimensional case, assuming a linear potential distribution between neighbouring grid points

$$\Psi(x) = \Psi_i + (\Psi_{i+1} - \Psi_i) \frac{x - x_i}{x_{i+1} - x_i} , \qquad \forall x \in [x_i, x_{i+1}] \qquad (3.45)$$

the finite difference equation becomes

Spatial Discretization

$$\frac{\Psi_{i+1}-\Psi_i}{x_{i+1}-x_i} - \frac{\Psi_i-\Psi_{i-1}}{x_i-x_{i-1}} = -\frac{1}{\varepsilon}\int_{\frac{(x_{i-1}+x_i)}{2}}^{\frac{(x_i+x_{i+1})}{2}} \rho(x)\,dx \ . \tag{3.46}$$

The final result depends on the approximation for $\rho(x)$. Assuming a constant space-charge density yields the classical one-dimensional finite difference form of Poisson's equation

$$\frac{\Psi_{i+1}-\Psi_i}{x_{i+1}-x_i} - \frac{\Psi_i-\Psi_{i-1}}{x_i-x_{i-1}} = -\frac{1}{2\varepsilon}(x_{i+1}-x_{i-1})\rho_i \ . \tag{3.47}$$

In the finite element method, the domain is likewise discretized into subdomains. In the one-dimensional case, a subdomain or an element is simply the interval $[x_i, x_{i+1}]$ between two neighbouring grid points. For the example considered here the same local approximation (3.45) yields the shape functions

$$u_i(x) = \begin{cases} \frac{x-x_{i-1}}{x_i-x_{i-1}} , & \forall x \in [x_{i-1}, x_i] \\ \frac{x_{i+1}-x}{x_{i+1}-x_i} , & \forall x \in [x_i, x_{i+1}] \\ 0 , & \forall x \notin [x_{i-1}, x_{i+1}] \end{cases} \tag{3.48}$$

which represent a basis for the global approximation of the unknown potential

$$\Psi(x) = \sum_i \Psi_i u_i(x) \ . \tag{3.49}$$

The finite element equations can now be obtained for Poisson's equation by applying either Galerkin's method directly to the differential equation or the Ritz method to an equivalent variational problem. Applying Galerkin's procedure to Poisson's equation gives

Semiconductor Device Simulation

$$\int_v (\nabla \cdot \nabla \Psi - \rho) u_i dv = 0 \tag{3.50}$$

where all dependent variables are approximated via first-order polynomial shape functions as in (3.48) and (3.49). The following transformation is then made

$$\int_v (\nabla \cdot \nabla \Psi - \rho) u_i dv = \int_{dv} u_i \cdot \nabla \Psi \cdot dl - \int_v (\nabla \Psi \cdot \nabla u_i - \rho u_i) dv \tag{3.51}$$

The first term on the right hand side of (3.51) vanishes in the one-dimensional case resulting in

$$\int_v \nabla u_i \cdot \nabla \Psi dv = -\frac{1}{\varepsilon} \int u_i \rho dv . \tag{3.52}$$

The integration domain covers only those elements which are incident with the considered node i, because u_i and ∇u_i vanish elsewhere. Evaluating (3.52) with the approximations (3.48) and (3.49) yields

$$\frac{\Psi_{i+1} - \Psi_i}{x_{i+1} - x_i} - \frac{\Psi_i - \Psi_{i-1}}{x_i - x_{i-1}} = -\frac{1}{\varepsilon} \int_{x_{i-1}}^{x_{i+1}} u_i(x) \rho(x) dx . \tag{3.53}$$

This result differs from the finite difference equation (3.46) only slightly in the way that space-charge density is taken into account at both right-hand side integrals. In (3.46), $\rho(x)$ is directly integrated over the domain of a box, whereas the finite-element approach requires a weighted integration over the larger domain covering adjacent domains. Moreover, both methods yield the difference equation (3.47), if a constant space-charge density is assumed within the integration intervals.

From the above considerations for Poisson's equation in one dimension one may conclude that nearly identical difference equations are obtained from the finite difference and finite element methods if in both methods similar approximations are used. This is even true if more than one dimension is considered. For a node (x_i, y_j), the rectangular grid shown in Figure 3.2, the standard difference formulation of Poisson's equation can be computed to be

Spatial Discretization

$$\frac{1}{2}(y_{j+1}-y_{j-1})\left[\frac{\Psi_{i+1,j}-\Psi_{i,j}}{x_{i+1}-x_i}-\frac{\Psi_{i,j}-\Psi_{i-1,j}}{x_i-x_{i-1}}\right]$$

$$+\frac{1}{2}(x_{i+1}-x_{i-1})\left[\frac{\Psi_{i,j+1}-\Psi_{i,j}}{y_{j+1}-y_j}-\frac{\Psi_{i,j}-\Psi_{i,j-1}}{y_j-y_{j-1}}\right]$$

$$=\frac{1}{4\varepsilon}(x_{i+1}-x_{i-1})(y_{j+1}-y_{j-1})\rho_{i,j}$$

(3.54)

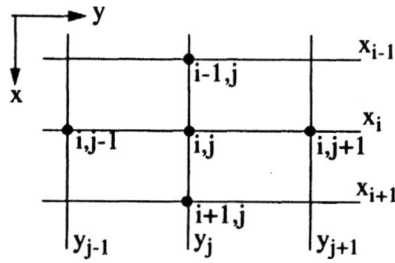

FIGURE 3.2 Cell of a rectangular grid.

This equation can be obtained using either method if the appropriate approximations are used. It is widely accepted that the major advantage of finite element methods becomes apparent if triangular grids are used, since triangular elements make local refinements more readily available without excessively increasing the total number of nodes. Depending on the specific example, the difference in the number of grid points can be much larger than for the simple example shown in Figure 3.3. Moreover, non-planar bounaries or interfaces can be more precisely approximated by triangles than rectangles. On the other hand, such problems can also be solved with finite difference methods by introducing triangular subregions within the rectangles [3.30] or more generally, by using directly a triangular grid [3.29],[3.31],[3.2]. If in the latter case the box integration method is applied, the perpendicular bisectors of the lines between neighbouring grid points define the box associated with a grid point as indicated in Figure 3.4. Usually the potential of triangular grids is not fully exploited because this demands an increased effort for mesh generation and book keeping. This is even more pronounced when three-dimensional tetrahedral elements are used [3.32].

Semiconductor Device Simulation

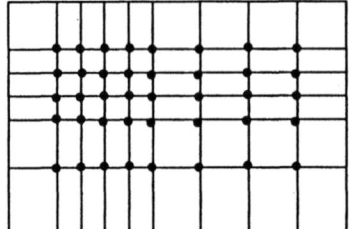 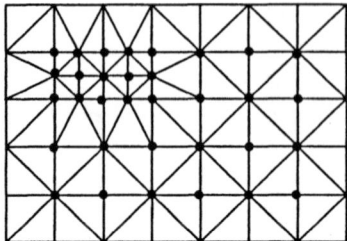

FIGURE 3.3 Local refinement in a rectangular and triangular grid.

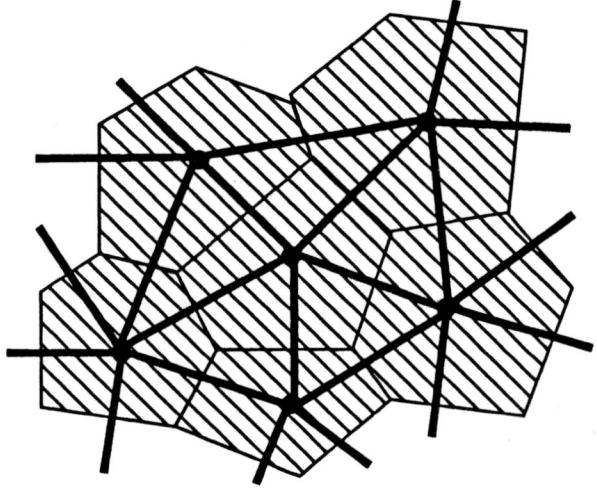

FIGURE 3.4 Triangular grid and the boxes associated with them.

Of equal importance as matching the grid to the geometrical constrainst of the problem, is the adaptation of the approximate solutions to the physics of the problem, in particular if the properties of the solutions are dominated by nonlinearities. In other words, trial functions to be used should primarily reflect the physics involved, whether they are simultaneously shape functions or not. Hence, trial functions which use approximate analytical solutions of the problem are a good choice.

3.5.2 Scharfetter-Gummel Transformation

The key mathematical technique in device simulation is the formulation of the current equations proposed by Scharfetter and Gummel [3.33]. The basic problem is that the standard drift-diffusion form of the current density equations (3.6),(3.7) in a semiconductor is numerically unstable if the carrier concentrations are assumed to vary linearly between mesh points. In particular, at p-n junctions and semiconductor interfaces the carrier concentrations can vary by twenty orders of magnitude in a few tenths of a micron. Consequently, such a large variation cannot be modeled well by a low order approximating function. To achieve a stable solution, Scharfetter and Gummel proposed a simple solution as follows.

Assuming steady-state conditions, the continuity equation for holes (3.7) can be written as

$$\frac{1}{q}\nabla J_p = -\nabla\left(\mu_p n_{ie} exp\left(\frac{\phi_p - \Psi}{V_T}\right)\nabla\phi_p\right) = -R \tag{3.55}$$

where (3.9) and (3.29) have been used. Neglecting recombination within the one-dimensional interval $[x_i, x_{i+1}]$ and integrating (3.55) yields a constant current density in the interval, $J_{p\ i+1/2}$

$$\frac{J_p(x)}{q} = \frac{J_{pi+\frac{1}{2}}}{q} = -\frac{\zeta_p(x) - \zeta_{pi}}{\int_{x_i}^{x} \frac{exp\left(\frac{\Psi}{V_T}\right)}{V_T \mu_p n_{ie}}dx'} = \frac{\zeta_{pi+1} - \zeta_{pi}}{\int_{x_i}^{x_{i+1}} \frac{exp\left(\frac{\Psi}{V_T}\right)}{V_T \mu_p n_{ie}}dx'} = const \tag{3.56}$$

where $\zeta_p = exp\left(\frac{\phi_p}{V_T}\right)$.

Because the interval considered is only small, the linear potential distribution (3.45) and a constant product of mobility and effective intrinsic density $(\mu_p, n_{ie})_{i+\frac{1}{2}}$ can be assumed there. Substituting (3.45) for Ψ in (3.56) yields

$$\zeta_p(x) = \zeta_{pi} + (\zeta_{pi+1} - \zeta_p) \cdot \frac{exp\left(\frac{\Psi(x)}{V_T}\right) - exp\left(\frac{\Psi_i}{V_T}\right)}{exp\left(\frac{\Psi_{i+1}}{V_T}\right) - exp\left(\frac{\Psi_i}{V_T}\right)} \qquad (3.57)$$

and

$$\frac{J_{pi+\frac{1}{2}}}{q} = -(\mu_p n_{ie})_{i+\frac{1}{2}} \cdot \frac{\Psi_{i+1} - \Psi_i}{x_{i+1} - x_i} \cdot \frac{\zeta_{pi+1} - \zeta_{pi}}{exp\left(\frac{\Psi_{i+1}}{V_T}\right) - exp\left(\frac{\Psi_i}{V_T}\right)} \qquad (3.58)$$

Although the original formulation of this discretization scheme by Scharfetter and Gummel employed the finite difference approach, comparing the approximate solution of the continuity equation with the general expression for a global approximation of the unknown parameter ζ_p

$$\zeta_p(x) = \sum_i \zeta_{pi} v_i(x) \qquad (3.59)$$

we see that the basis function $v_i(x)$ for a finite element representation of the continuity equations can be defined as follows [3.34]

$$v_i(x) = \begin{cases} \dfrac{exp\,(\Psi(x) - \Psi_{i-1})}{exp\,(\Psi_i - \Psi_{i-1})} & , \quad \forall x \in [x_{i-1}, x_i] \\[2mm] \dfrac{exp\,(\Psi_{i+1} - \Psi(x))}{exp\,(\Psi_{i+1} - \Psi_i)} & , \quad \forall x \in [x_i, x_{i+1}] \\[2mm] 0 & , \quad \forall x \notin [x_{i-1}, x_{i+1}] \end{cases} \qquad (3.60)$$

Substituting (3.58) into the finite difference equation

Spatial Discretization

$$\frac{1}{q}\left(J_{pi+\frac{1}{2}} - J_{pi-\frac{1}{2}}\right) = -\frac{1}{2}(x_{i+1} - x_{i-1}) R(x_i) \tag{3.61}$$

results in a highly efficient approximation for the one-dimensional continuity equation.

The discretization scheme mentioned above has generally a much larger range of validity than more conventional ones, where the current densities are derived from polynomial trial functions for the quasi-Fermi level or the current density. Equation (3.58) agrees with more conventional expressions only if in the latter it is assumed that $|\Psi_{i+1} - \Psi_i| \ll V_T$ and $|\phi_{pi+1} - \phi_{pi}| \ll V_T$.

Since these assumptions are not made here, the finite difference approximation (3.58), (3.61) improves numerical accuracy or enables the use of coarse grids in regions where the assumptions of the analytical integral (3.56) are not strongly violated. Gummel's discretization scheme can also be extended to the 2D case. In this case the current density components along mesh lines are assumed to be constant between neighbouring grid points and approximated by an equation of the form (3.58).

Spatial discretization of the field equations yields at each grid point one discrete equation for Poisson's equation and both continuity equations, respectively. The resulting system comprises ordinary differential equations in the time domain because the spatially discretized continuity equations still contain derivatives with respect to time. Condensing the nodal values of the unknown functions into vectors, this system is written as

$$f_\Psi(\Psi, \phi_n, \phi_p, t) = 0 \tag{3.62}$$

$$f_n(\Psi, \phi_n, \phi_p, t) = \frac{\partial n}{\partial t} \tag{3.63}$$

$$f_p(\Psi, \phi_n, \phi_p, t) = \frac{\partial p}{\partial t} \tag{3.64}$$

where the vector functions f_Ψ, f_n and f_p represent the spatial difference approximations of the field equations.

Integration of the system (3.62)-(3.64) requires again numerical algorithms discretizing the time domain and yielding difference approximations for the derivatives. Though semi implicit methods can also be used, usually fully implicit backward dif-

Semiconductor Device Simulation

ference formulas are applied, in order to guarantee nuerical stability without the imposition of small time steps [3.36]. In most cases, the first-order backward Euler formula is used [3.37]-[3.40], since it is "A-stable" for constant time steps and for increasing step size with a bounded rate of change [3.41]. This formula yields for (3.63)

$$f_n\left(\Psi^m, \phi_n^m, \phi_p^m, t^m\right) = \frac{n^m - n^{m-1}}{t^m - t^{m-1}} \quad (3.65)$$

where the superscripts denote the values of the corresponding values at time t^m. Moreover, implicit backward-difference formulas with variable step size and order of integration [3.42] can be profitably used [3.43],[3.44] ensuring accuracy as well as stability by an automatic control of steps and order [3.41]. Note, that for a fully implicit integration formula (e.g., (3.65)) the left hand side function has to be evaluated at time t^m. This requires for each time step the solution of a system of coupled equations including Poisson's equation (3.63).

3.6 Solution Methods

In a transient analysis, space discretization and implicit time integration yield at each instant of time the coupled nonlinear system

$$F_\Psi(\Psi, \phi_n, \phi_p) = 0 \quad (3.66)$$

$$F_n(\Psi, \phi_n, \phi_p) = 0 \quad (3.67)$$

$$F_p(\Psi, \phi_n, \phi_p) = 0 \quad (3.68)$$

where the argument t^m and the superscripts m are omitted for simplicity. In a steady-state analysis, a system of the same form is obtained for each bias point. Thus in both cases $3N$ nonlinear equations have to be solved, where, e.g., in the two dimensional case the number of grid points N can vary anywhere between 3000 and 10,000. The solution of this large system is the most time consuming task during a numerical device simulation. Therefore, it is of great importance to choose an efficient numerical algorithm, especially if more than one dimension has to be considered.

Most simply, one treats each of the differential equations by separately by decoupling the equations and solving systems (3.66)-(3.68) successively. First Poisson's equation

Solution Methods

(3.66) is solved assuming known quasi-Fermi levels. Next comes each of the continuity equations with Ψ given from the first step. This sequence is iteratively repeated until self-consistent values of the desired accuracy for all unknown variables are obtained. Figure 3.5 shows the flow diagram of this successive or decoupled method, which is due to Gummel [3.45]. It is advantageous for multidimensional simulations as it saves storage and converges quite well as long as the coupling of th three equations is only weak. Moreover, the result of the first cycle of this iteration is rather close to the exact solution if the first guess of the quasi-Fermi potential of the majority carriers is accurate enough.

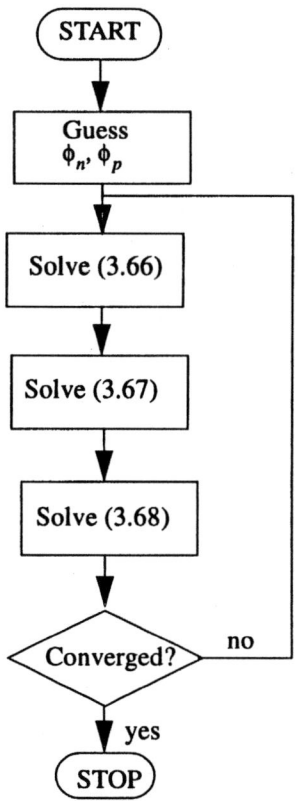

FIGURE 3.5 Gummel's algorithm.

Semiconductor Device Simulation

Since each of the three systems (3.66)-(3.68) which have to be solved one after the other within Gummel's algorithm, is generally nonlinear, an iterative method has to be used. Because of its quadratic convergence, the Newton method is usually chosen. Applying this method, e.g., to Poisson's equation, the system (3.66) is linearized and the iteration is defined as follows:

$$\frac{\partial F_\Psi}{\partial \Psi} \delta \Psi_{k+1} = -F_\Psi(\Psi_k) \tag{3.69}$$

$$\Psi_{k+1} = \Psi_k + \delta \Psi_{k+1} \qquad k = 0, 1, 2, \ldots \tag{3.70}$$

starting with an initial approximation Ψ_0. $\partial F_\Psi / \partial \Psi$ is the Jacobian matrix evaluated at Ψ_k and $\delta \Psi_{k+1}$ is the correction vector at the iteration step $k+1$.

The relatively simple implementation of the successive method has to be paid for by its possible slow convergence if (3.66)-(3.68) are strongly coupled. Considering the steady-state analysis of a bipolar device, Gummel's algorithm converges well for low and moderate injection levels. But if for higher forward voltages the minority carrier density approaches the majority carrier density, both carriers become strongly coupled via the electrostatic potential in order to maintain quasi-neutrality and convergence slows down rapidly [3.46]. For a FET analysis in the strong inversion region, the mutual coupling between the potential and the dominating carrier density is likewise increased [3.47] though originating from a different mechanism. Also, in transient simulations, the displacement current can give rise to additional coupling [3.39].

These convergence problems can be overcome mathematically by solving (3.66)-(3.68) simultaneously rather than alternatingly. This simultaneous method has the advantage that the mutual coupling between all equations is taken into account by a quadratically converging overall Newton iteration

$$\begin{bmatrix} \frac{\partial F_\Psi}{\partial \Psi} & \frac{\partial F_\Psi}{\partial \phi_n} & \frac{\partial F_\Psi}{\partial \phi_p} \\ \frac{\partial F_n}{\partial \Psi} & \frac{\partial F_n}{\partial \phi_n} & \frac{\partial F_n}{\partial \phi_p} \\ \frac{\partial F_p}{\partial \Psi} & \frac{\partial F_p}{\partial \phi_n} & \frac{\partial F_p}{\partial \phi_p} \end{bmatrix} \begin{bmatrix} \delta \Psi_{k+1} \\ \delta \phi_{n,k+1} \\ \delta \phi_{p,k+1} \end{bmatrix} = \begin{bmatrix} -F_{\Psi,k} \\ -F_{n,k} \\ -F_{p,k} \end{bmatrix} . \tag{3.71}$$

Although this simultaneous Newton method is advantageous from a purely mathematical viewpoint, it is more involved with regard to program structure and storage requirements. Hence, comparisons between both methods [3.47] do not only depend on the device and its operating conditions but, in addition, also on the algorithms implemented for solving the different linearized systems.

For one-dimensional simulations the simultaneous method is usually applied [3.38] amd the linear system (3.71) is best solved by Gaussian elimination or LU decomposition. An appropriate ordering of the columns and rows of the Jacobian yields a simple band matrix with mostly seven nonzero diagonals [3.44] allowing the efficient use of special band solver routines. For two-dimensional problems, this method proves to be also successfull [3.48], provided that the bandwith of the Jacobian matrix is not too large.

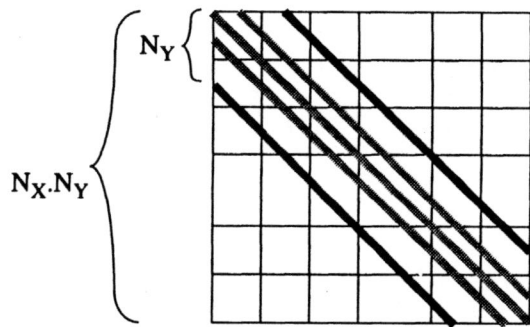

FIGURE 3.6 Five-diagonal band matrix. N_x and N_y are the numbers of grid lines in the x-direction and y-direction respectively.

In Figure 3.6, a five-diagonal band matrix is depicted, as it arises in the decoupled method from (3.68) assuming, e.g., the difference equation (3.54) and a rectangular grid of N_x and N_y grid lines in the x- and y- direction, respectively. For a direct elimination process, computation times and storage requirements are proportional to $N_y^3 N_x$ and $N_y^2 N_x$, respectively, if one takes into account the fill-ins of nonzero coefficients within the bandwidth N_y. Threefold size and bandwidth of the Jacobian (3.71) for the simultaneous procedure leads to an increase of storage and computation time by a factor of nine compared with one Newton step for each decoupled equation. Hence this approach is prohibitive unless large computer resources are available [3.49].

Semiconductor Device Simulation

If an irregular grid (as shown in Figure 3.3 or Figure 3.4) is used, both Jacobian matrices (3.70),(3.71) are still sparse, but they lose their simple band structure, yet allowing the application of direct elimination methods if the increase of programming effort and program complexity is accepted. Such grids require pre-processing for symbolic LU factorization and ordering of the sparse matrix equation s [3.50],[3.40]. Quite often restricted computer resources require an additional internal iteration for the solution of the large systems of linearized equations. This is especially true if the simultaneous method has to be used for two- and three-dimensional simulations. Various numerical algorithms can be successfully applied to the linear equations, e.g., Stone's strongly implicit procedure (SIP) [3.51],[3.52], the successive line over relaxation (SLOR) method [3.53],[3.46],[3.54] or the Incomplete Choleski Conjugate Gradient (ICCG) method [3.55],[3.56]. The main disadvantage of these iterative methods is their linear convergence which is often rather slow and can limit considerably the available accuracy of the solution.

One common approach to escape convergence problems is to start at a solution for the device in thermal equilibrium and then try to climb up the operating dc characteristics to the requested bias point step by step using the solution of the preceding bias point as an initial guess to solve for the subsequent bias point. Since in many cases one is interested in a set of solutions for a number of bias points anyhow, this approach seems to be quite obvious. Although it is assured to stay within the radius of convergence for the Newton method, it fails to converge in the simpler Gummel procedure when coupling between the respective equations becomes strong. In some device simulators [3.2], initial guesses are made locally. This takes the solution in memory, sets the applied bias, and changes the majority carrier Fermi potentials throughout heavily-doped regions to be equal to the bias applied to that region. This procedure is effective in the context of a Gummel iteration, particularly in reverse bias, but less so for a Newton method.

Broadly speaking, Newton's method with Gaussian elimination of the Jacobian is by far the most stable method of solution. Unfortunately it can be expensive for two-carrier simulations, both in time and memory. For low current solutions the Gummel method offers an attractive alternative to inverting the full Jacobian.

3.7 A Representative Example

In the past, numerical solutions of the basic equations in one, two or three space dimensions have been obtained for several different devices in a large variety of configurations. These numerical models permit one to look into an operating device and watch the different internal mechanisms. By this, one can trace the dominating effects in the respective device regions and sharpen one's physical understanding of electrical device behaviour. The knowledge gained during the process can quite often be valued higher than the numerical results themselves, since it may guide the course for improved design. Since in this book we are concerned with the problems of substrate coupling and solutions for it we will use a device simulator, MEDICI [3.2] to simulate a set up representative of the problem.

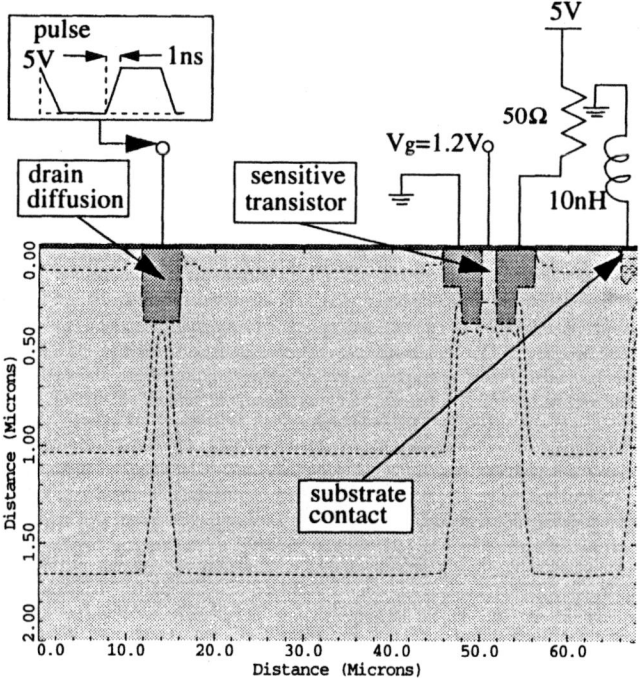

FIGURE 3.7 A representative example of the substrate coupling problem.

Semiconductor Device Simulation

TABLE 3.3 **Features of the 2 um BiCMOS technology**

substrate: p-ep/p+ wafer

well: n-well

n-well depth: 4.9 µm

epitaxial layer thickness: 16 µm

epitaxial layer doping: 9×10^{14} /cm^3

gate oxide thickness: 24 nm

threshold voltage (NMOS): 0.51V

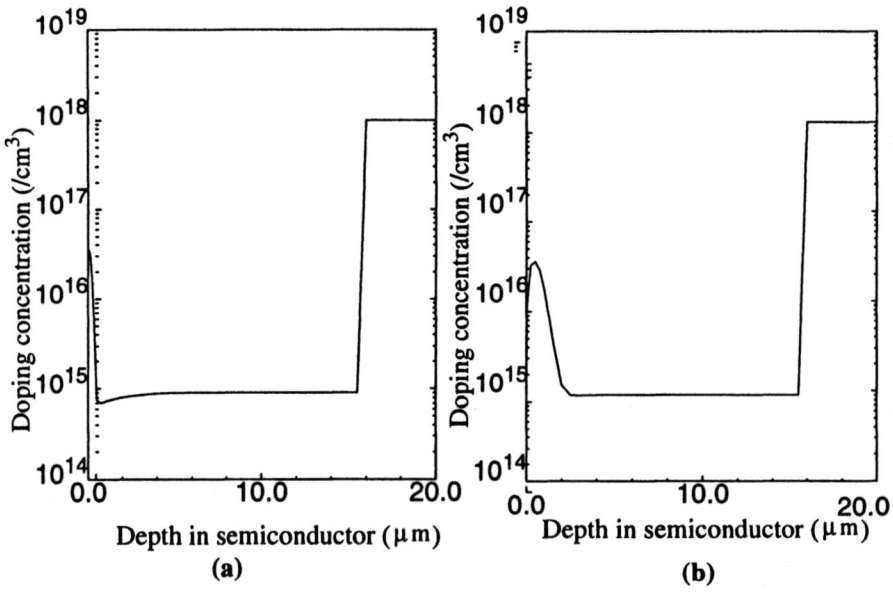

FIGURE 3.8 **Impurity doping concentration versus depth in semiconductor. a) at the NMOS transistor gate and b) under the field oxide.**

A Representative Example

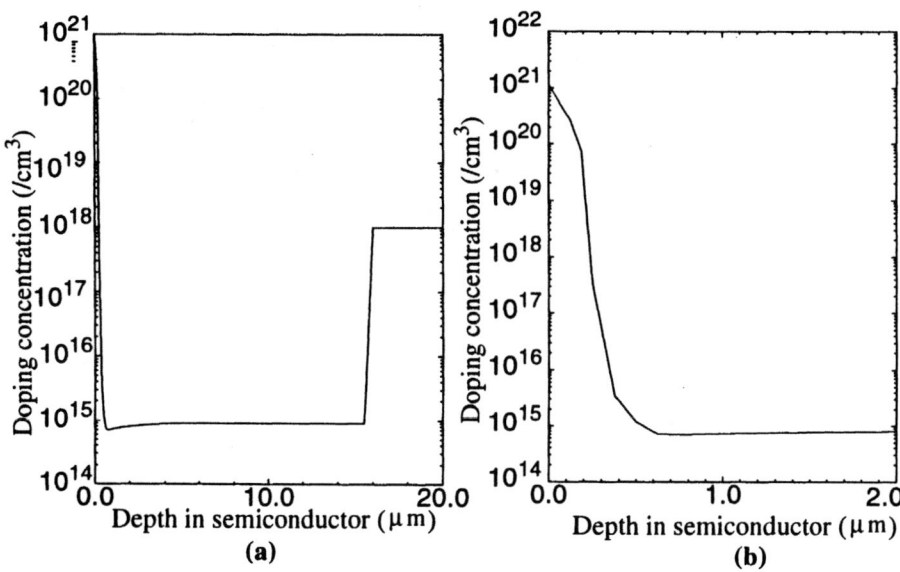

FIGURE 3.9 Impurity doping concentration versus depth in semiconductor. a) at the NMOS transistor drain and b) magnified view.

The semiconductor is specified by means of a doping profile as shown in Figure 3.7. The lower part of this figure corresponds to a device structure, that includes a diffused region equivalent to the drain of a switching digital transistor, an NMOS transistor considered to be part of a noise sensitive circuit and a substrate contact to bias up the p type substrate. The doping profile of a 2 μm BiCMOS process [3.57] has been used for the semiconductor and some of its key features are listed in Table 3.3 [3.58]. One dimensional plots of the impurity doping concentrations versus depth in the semiconductor at the gate, drain and under the field oxide in the p substrate for this process are shown in Figure 3.8 and Figure 3.9.

Besides its device simulation capability MEDICI also provides the user the opportunity to simulate lumped circuit elements in conjunction with the semiconductor model. In essence, this allows for circuit simulation with accurate numerical models for the semiconductor devices. The circuit elements must be connected to the boundaries of the semiconductor being simulated.

The upper part of the diagram in Figure 3.7 includes the supply voltages, a CMOS level excitation pulse with 1 ns transition times applied to the equivalent drain diffu-

Semiconductor Device Simulation

sion and a couple of lumped circuit elements. The parasitic inductor models the package (bond wire and package pin)while the 50Ω resistor is used to represent the input impedance of the test instrumentation.

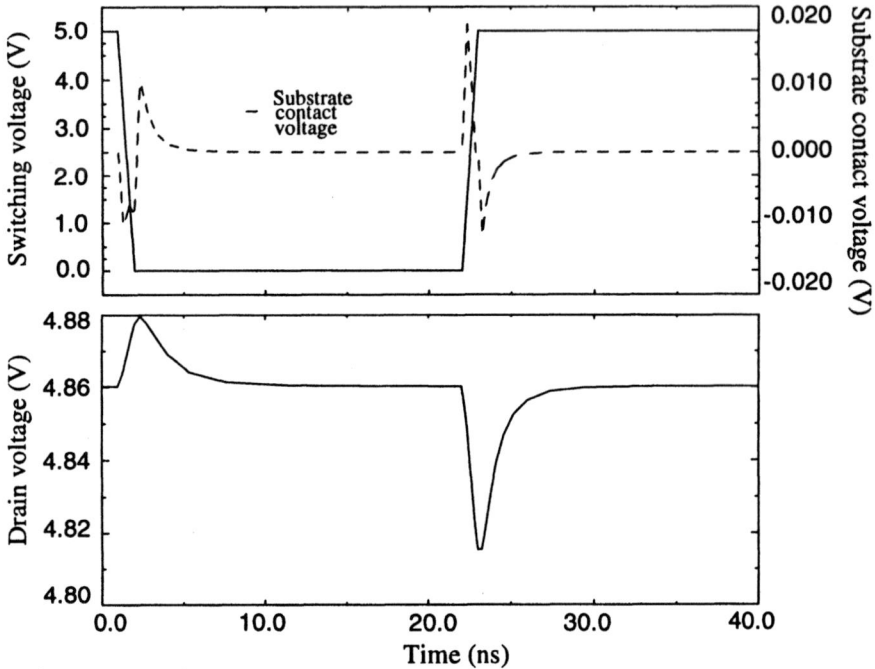

FIGURE 3.10 Transient simulation result showing the substrate coupling effect on the drain node of the sensitive transistor in Figure 3.7.

As the equivalent drain of the digital transistor switches from 5V to 0V, the substrate potential under this drain experiences a fluctuation because of the drain junction capacitance. Initially the substrate potential spikes down as the derivative of the switching voltage decreases (the switching voltage falls from its initial value of 5V) and then in the opposite direction as the switching voltage levels off to attain its final value of 0V. These fluctualtions in the substrate voltage spead across the substrate terminating at the contact which first sources and then sinks current in an attempt to return to an equilibrium state. The flow of current across the substrate results in similar fluctuations in the body potential of the sensitive transistor which is in the path of the flow. As the body voltage of this transistor falls, owing to the body effect, its threshold inreases. Hence, the drain current decreases causing the drain voltage to

Summary

increase ($V_{DD} - I_D.R$). The drain voltage continues to increase until the body voltage starts to rise after which it decreases until it reaches its equilibrium value. Similarly, a negative pulse appears at the drain of the sensitive transistor when the equivalent drain of the digital transistor switches from 0V back to 5V. Figure 3.10 shows the switching pulse, the resulting voltage waveform at the substrate contact and its effect on the drain voltage of the sensitive transistor.

While MEDICI computes the effects of substrate coupling accurately, it is primarily a device simulator to be used to investigate device behaviour. Its feasibility in simulating substrate coupling is limited to a very few transistors and thus cannot be extended to simulate these effects on a real chip with several thousands let alone hundreds of thousands of transistors. To address the latter problem, simplified solution procedures are required and these will be discussed in the next few chapters.

3.8 Summary

In this chapter we reviewed the various numerical techniques, modeling and simulation strategies employed in standard device-level simulators. A popular device simulation program, MEDICI was employed to study the nature of substrate coupling in a representative example.

REFERENCES

[3.1]   W.L. Engl, H.K. Dirks, B. Meinerzhagen, "Device Modeling," *Proceedings of the IEEE*, Vol. 71, No. 1, Jan. 1983.

[3.2]   *TMA MEDICI : Two Dimensional Device Simulation Program*, Version 1, Volume 1, Technology Modeling Associates, Inc., 1992.

[3.3]   J.B. Johnson, *The Voronoi Cell Method for Two and Three Dimensional Semiconductor Device Simulation*, Ph.D dissertation, Carnegie Mellon University, 1987.

[3.4]   W.L. Engl, R. Laur and H.K. Dirks, "MEDUSA- A Simulator for Modular Circuits," *IEEE Transactions on Computer-Aided Design of Integrated Circuits*, Vol. CAD-1, pp. 85-93, 1982.

[3.5] O. Madelung, *Introduction to Solid-State Theory*, Berlin: Springer, 1978.

[3.6] C.R. Crowell and S.M. Sze, "Current transport in metal-semiconductor barriers," *Solid State Electronics*, vol. 9, pp. 1035, 1966.

[3.7] D. M. Caughey and R.E. Thomas, "Carrier mobilities in silicon empirically related to doping and field," *Proceedings of the IEEE*, vol. 55, pp. 2192-2193, 1967.

[3.8] S. Selberherr, "Process and device modeling for VLSI," *Microelectronics Reliability*, vol. 24, no. 2, pp. 225-257, 1984.

[3.9] S. M. Sze, *Physics of Semiconductor Devices*, 2nd ed., John Wiley and Sons, New York, 1981.

[3.10] Zhiping Yu and R.W. Dutton, *SEDAN III - A Generalized Electron Material Device Analysis Program*, Stanford Electronics Laboratory Technical Report, Stanford University, July 1985.

[3.11] F. Dannhauser, "Die Abhangigkeit der Tragerbeweglichkeit in Silizium von der Konzentration der freien Ladungstrager-I," *Solid-State Electronics*, vol. 15, pp. 1371-1375, 1972.

[3.12] J. Krausse, "Die Abhangigkeit der Tragergeweglichkeit in Silizium von der Konzentration der freien Ladungstrager- II," *Solid-State Electronics*, vol. 15, pp. 1377-1381, 1972.

[3.13] W. Anheier and W.L. Engl, "Numerical analysis of gate triggered SCR turn-on transients," in *Technical Digest of the IEEE International Electron Devices Meeting*, pp. 303A-303D, 1977.

[3.14] K. K. Thornber, "Relation of drift velocity to low-field mobility and high-field saturation velocity," *Journal of Applied Physics*, vol. 51, no. 4, pp. 594-602, 1970.

[3.15] K. Yamaguchi, "Field-dependent mobility model for two-dimensional numerical analysis of MOSFETs," *IEEE Transactions on Electron Devices*, vol. ED-26, pp. 1068-1074, 1979.

[3.16] J. Cooper and D.F. Nelson, "Measurement of the high-field velocity of electrons in inversion layers on silicon," *IEEE Electron Devices Letters*, vol. EDL-2, pp. 171-173, 1981.

[3.17] A.H. Marshak and K.M. van Vliet, "Electric current in solids with position-dependent band structure," *Solid-State Electronics*, vol. 21, pp. 417-427, 1978.

[3.18] A.H. Marshak and K.M. van Vliet, "Carrier densities and emitter efficiency in degenerate materials with position-dependent band structure," *Solid-State Electronics*, vol. 21, pp. 429-434, 1978.

[3.19] J.W. Slotboom and H.C. De Graaff, "Measurements of bandgap narrowing in Si bipolar transistors," Solid-State Electronics, vol. 19, pp. 857-862, 1976.

[3.20] A.W. Wieder, "Emitter effects in shallow bipolar devices: Measurements and consequences," *IEEE Transactions on Electron Devices*, vol. ED-27, pp. 1402-1408, 1980.

[3.21] R.P. Mertens, J.L. van Meerbergen, J.F. Nijs, and R.T. van Overstraeten, "Measurement of the minority-carrier transport parameters in heavily doped silicon," *IEEE Transactions on Electron Devices*, vol. ED-27, no. 5, 1980.

[3.22] A. Selloni and S.T. Pantelides, *Phys. Rev. Lett.*, Oct. 15, 1982.

[3.23] D.L. Scharfetter, "Measured dependence of lifetime upon defect density and temperature in depletion layers of epitaxial silicon diodes," presented at the Solid-State Devices Res. Conference, Santa Barbara, CA, 1967.

[3.24] J.G. Fossum and D.S. Lee, "A physical model for the dependence of carrier lifetime on doping density in nondegenerate silicon," *Soid-State Electronics*, vol. 25, no. 8, pp. 741-747, 1982.

[3.25] J. Dziewor and W. Schmid, "Auger coefficients for highly doped and highly excited silicon," *Applied Physics Letters*, vol. 31, pp. 346-348, 1977.

[3.26] J.O. Beck and R. Conradt, "Auger recombination in silicon," *Solid-State Communications*, vol. 13, pp. 93-95, 1973.

[3.27] A.S. Grove, *Physics and Technology of Semiconductor Devices*, John Wiley and Sons, New York, 1967.

[3.28] G. Strang and G.J. Fix, *An Analysis of the Finite Element Method*, Englewood Cliffs, NJ: Prentice-Hall, 1973.

[3.29] R.S. Varga, *Matrix Iterative Analysis*, Englewood Cliffs, NJ: Prentice-Hall, 1962.

[3.30] J.A. Greenfield and R.W. Dutton, "Nonplanar VLSI device analysis using the solution of Poisson's equation," *IEEE Transactions on Electron Devices*, vol. ED-27, pp. 1520-1532, 1980.

[3.31] R.H. Mac Neal, "An asymmetrical finite difference network," *Quart. Appl. Math.*, vol. 11, pp. 295-310, 1953.

[3.32] E.M. Buturla, P.E. Cottrel, B.M. Grossman, and K.A. Salsburg, "Finite-element analysis of semiconductor devices: The Fieldday program," *IBM Journal of Research and Development*, vol. 25, pp. 218-231, 1981.

[3.33] D.L. Scharfetter and H.K. Gummel, "Large-Signal Analysis of a Silicon Read Diode Oscillator'" *IEEE Transactions on Electron Devices*, vol. 16, no. 1, pp. 64-77, 1969.

[3.34] W.L. Engl and H. Dirks, "Numerical Device Simulation Guided by Physical Approaches," in *Numerical Analysis of Semiconductor Devices*, B.T. Browne and J.J.H. Miller, Eds., Dublin, Ireland: Boole Press, 1979.

[3.35] M.S. Mock, "Time-dependent simulation of coupled devices," in *Numerical Analysis of Semiconductor Devices and Integrated Circuits*, B.T. Browne and J.J.H. Miller, Eds., Dublin, Ireland: Boole Press, 1981.

[3.36] M.S. Mock, "Time discretisation of a nonlinear initial value problem," *Journal of Computational Physics*, vol. 21, pp. 20-37, 1976.

[3.37] A. De Mari, "An accurate numerical one-dimensional solution of the p-n junction under arbitrary transient conditions," *Solid-State Electronics*, vol. 11, pp. 1021-1053, 1968.

[3.38] G.D. Hachtel, R.C. Joy, and J.W. Cooley, "A new efficient one-dimensional analysis program for junction device modeling," *Proceedings of the IEEE*, vol. 60, pp. 86-98, 1972.

[3.39] O. Manck and W.L. Engl, "Two dimensional computer simulation for switching a bipolar transistor out of saturation," *IEEE Transactions on Electron Devices*, vol. ED-22, pp. 339-347, 1975.

[3.40] G.D. Hachtel, M.H. Mack, R.R. O'Brien and B. Speelpenning, "Semiconductor analysis using finite elements-Part I: Computational aspects," *IBM Journal of Research and Development*, vol. 25, pp. 232-245, 1981.

[3.41] H.P. Strohband, "New results on stability of the BDG-integration method with non-constant stepsize and order," in *Asilomar Conference on Circuits, Systems and Computers, Conference Records*, pp. 354-358, 1975.

[3.42] R.K. Brayton, F.G, Gustavson, and G.D. Hachtel, "A new efficient algorithm for solving differential-algebraic systems using implicit backward differnetiation formulas'" *Proceedings of the IEEE* , vol. 60, pp. 98-108, 1975.

[3.43] R. Laur and J.P. Strohband, "Numerical modeling technique for computer-aided circuit design," in *Proceedings of the IEEE International Symposium on Circuits and Systems*, pp. 247-250, 1976.

[3.44] W.L. Engl and H. Dirks, "Functional device simulation by merging numerical building blocks," in *Numerical Analysis of Semiconductor Devices adn Integrated Circuits*, B.T. Browne and J.J.H. Miller, Eds: Dublin, Ireland: Boole Press, 1981.

[3.45] H.K. Gummel, "A self-consistent iterative scheme for one-dimensional steady-state transistor calculations," *IEEE Transactions on Electron Devices*, vol. ED-11, pp. 455-465, 1964.

[3.46] O. Manck, H.H. Heimeier, and W.L. Engl, "High injection in a two-dimensional transistor," *IEEE Transactions on Electron Devices*, vol. ED-21, pp. 403-409, 1974.

[3.47] E.M. Buturla and P.E. Cotrell, "Simulation of semiconductor transport using coupled and decoupled solution techniques," *Solid-State Electronics*, vol. 23, pp. 331-334, 1980.

[3.48] W.L. Engl, O. Manck and A.W. Wieder, "Modeling of bipolar devices," *in Process and Device Modeling for Integrated Circuits*, F. van de Wiele, W.L. Engl and P.G. Jespers, Eds., Leyden, The Netherlands: Noordhoff Int. Publ., pp. 377-418, 1977.

[3.49] J.D'Arcy, E.J. Prendergast and P. Loyd, "Modeling of bipolar device structures- Physical simulations," *Technical Dogest of the International Electron Device Meeting*, pp. 516-519, 1981.

[3.50] T. Adachi, A. Yishii, and T. Sudo, "Two-dimensional semiconductor analysis using finite-element method," *IEEE Transactions on Electron Devices*, vol. ED-26, pp. 1026-1031, 1979.

[3.51] H.H. Heimeier, "A two dimensional numerical analysis of a silicon n-p-n transistor," *IEEE Transactions on Electron Devices*, vol. ED-20, pp. 708-714, 1969.

[3.52] S. Selberherr, A. Schutz, and H.W. Potzl, "Minimos- A two dimensional MOS transistor analyzer," *IEEE Transactions on Electron Devices*, vol. ED-27, pp. 1540-1549, 1980.

[3.53] J.W. Slotboom, "Computer-aided two dimensional analysis of bipolar transistors," *IEEE Transactions on Electron Devices*, vol. ED-20, pp. 669-679, 1973.

[3.54] D. Vandorpe, J. Borel, G. Merckel, and P. Saintot, "An accurate two-dimensional numerical analysis of the MOS transistor," *Solid-State Electronics*, vol. 15, p. 547, 1972.

[3.55] A. Yoshii, S. Horiguchi, M. Tomizawa, H. Kitazawa, and T. Sudo, "Three-dimensional analysis for semiconductor devices," *Japan Inst. Electron. Comm. Enggineers Tech. Rep. SSD-80-15*, pp.55-62, 1980 (in Japanese).

[3.56] T. Wada and R.L.M. Dang, "Modification of ICCG method for application to semiconductor device simulators," *Electronics Letters*, vol. 18, pp. 256-266, 1982.

[3.57] *The Stanford BiCMOS Project Annual Report*, Center for Integrated Systems, Stanford Univeristy, pp. 7-24, 1990.

[3.58] S. Masui, "Simulation of Substrate Coupling in Mixed-Signal MOS Circuits," *Technical Digest of the VLSI Circuits Symposium*, pp. 42-43, June 1992.

CHAPTER 4
Simplified Substrate Modeling and Rapid Simulation

As discussed earlier a semiconductor device simulator is an accurate computer-aided design tool to investigate substrate coupling related behaviour in a mixed-signal environment, although to simulate circuits with more than a few transistors can require overly large amounts of computational time and resources. The use of such a simulator is useful in terms of studying the characteristics of coupling through the substrate in a given process and to obtain direction in terms of strategies to overcome it [4.3]. However, the amount of coupling is extremely layout dependent and can vary dramatically depending on the nature of the substrate biasing and guard ringing employed and also the distribution of the power supplies used to bias the substrate. Consequently, even though layout guidelines can be established and employed to reduce the coupling, it becomes necessary to verify real designs, ranging anywhere in size from a few transistors to hundreds of thousands of them in order to break the expensive cycle of design, layout and fabrication. Thus, mixed-signal designers require a tool that they can use in conjunction with a circuit simulator which will indicate to them signs, if any, of performance deterioration due to substrate noise. Not only does the tool need to be accurate, it must also be fast enough to allow them to use it as many times as is required to optimize a design to meet its performance specifications without sacrificing large amounts of design time.

In order to develop a verification tool that is accurate and fast we will begin by making some simplifications on the physical equations that govern semiconductor behavior. Since the strategy is to use the tool in conjunction with a circuit simulator and semiconductor devices are abstracted into multi-terminal elements, we only need to deal with the substrate outside of these devices.

4.1 Simplified Equation

We have seen that semiconductor behaviour can be characterized by Poisson's equation (3.2), and the continuity equations for electrons and holes (3.6) and (3.7) which are repeated here for clarity

$$\nabla^2 \Psi = -\frac{\rho}{\varepsilon} \quad (4.1)$$

$$\frac{\partial n}{\partial t} - \frac{1}{q} \nabla \cdot J_n = -R \quad (4.2)$$

$$\frac{\partial p}{\partial t} + \frac{1}{q} \nabla \cdot J_p = -R. \quad (4.3)$$

Outside the semiconductor device/active areas and the contact areas, the substrate can be approximated as layers of varying doping density [4.1],[4.2]. In these regions, current flow across a concentration gradient of free (mobile) carriers is negligible due to the homogeniety of the semiconductor. The diffusion component of the the current density equations (3.10) and (3.11) can therefore be neglected and the equations for current density can be rewritten as

$$J_n = -q\mu_n n \nabla \Psi \quad (4.4)$$

$$J_p = -q\mu_p p \nabla \Psi \quad (4.5)$$

i.e., current flow is due to drift associated with potential gradient $\nabla \Psi$ only. Hence

$$J_n + J_p = -q(\mu_n n + \mu_p p) \nabla \Psi \quad (4.6)$$

and since electric field intensity, $E = -\nabla V$

$$J_n + J_p = q(q\mu_n n + \mu_p p) E . \quad (4.7)$$

Simplified Equation

The relative contribution of the minority carrier term in the equation above can be ignored and the equation can be approximated for an n-type semiconductor as $J_n + J_p \approx q\mu_n n E$; Similarly for a p-type substrate, $J_n + J_p = q\mu_p p E$.

Assuming that the net accumulation of free majority carriers in the substrate, Δn (due to the difference in carrier numbers flowing in and out of the semiconductor and the recombination/generation processes within the semiconductor itself) is much less than the equilibrium concentration of majority carriers, n_0, a constant resistivity, ρ', independent of carrier flow, can be used to describe carrier flow in the substrate.

$$J_n + J_p = \frac{1}{\rho'} E \tag{4.8}$$

where

$$\rho' = \frac{1}{q\mu_n n_0} \tag{4.9}$$

and

$$n = n_0 + \Delta n. \tag{4.10}$$

Adding the two carrier transport equations (4.2) and (4.3), we get,

$$q\left(\frac{\partial p}{\partial t} - \frac{\partial n}{\partial t}\right) = -\nabla \bullet (J_n + J_p) = -\frac{1}{\rho'} \nabla \bullet E \tag{4.11}$$

Differentiating Poisson's equation (and substituting for ρ using (3.6) and (3.7))

$$\nabla \bullet E(t) = \frac{q}{\varepsilon}(p(t) - n(t) + N_D - N_A) \tag{4.12}$$

on both sides with respect to time yields

$$q\left(\frac{\partial p}{\partial t} - \frac{\partial n}{\partial t}\right) = \varepsilon \frac{\partial}{\partial t}(\nabla \bullet E). \tag{4.13}$$

From (4.12) and (4.13) we get,

$$\varepsilon \frac{\partial}{\partial t}(\nabla \bullet E) = -\frac{1}{\rho'} \nabla \bullet E \tag{4.14}$$

Simplified Substrate Modeling and Rapid Simulation

Simplified Substrate Modeling and Rapid Simulation

or,

$$\varepsilon \frac{\partial}{\partial t}(\nabla \cdot E) + \frac{1}{\rho'}\nabla \cdot E = 0. \quad (4.15)$$

Equation (4.15) is actually a simplified form of Maxwell's equations

$$\nabla \times H = J + \frac{\partial D}{\partial t} \quad (4.16)$$

$$\nabla \times E = -\frac{\partial B}{\partial t} \quad (4.17)$$

$$\nabla \cdot D = \rho \quad (4.18)$$

$$\nabla \cdot B = 0 \quad (4.19)$$

where H is the magnetic field strength, B the magnetic flux density, D the electric flux density and E the electric field intensity.

Since the wavelengths of the magnetic fields far exceed the dimensions of a typical die ($\lambda = 3 \times 10^8/v$ m where v is the maximum frequency of operation of the circuit), it is possible to neglect the effects of (4.17) on chip. Furthermore, (4.16) can be reduced using the identity $\nabla \cdot (\nabla \times a) = 0$ to give

$$\nabla \cdot (\nabla \times H) = \nabla \cdot J + \nabla \cdot \frac{\partial D}{\partial t} = 0. \quad (4.20)$$

Using the material constants, $D = \varepsilon E$ and $J = (1/\rho')E$, (4.20) reduces to (4.15).

4.2 Spatial Discretization

To solve (4.15) across the substrate, spatial discretization is employed. As discussed in Section 3.5 there are several possible choices in discretization methods each with its own pros and cons. While finite difference methods such as the box integration method are simple and easy to employ they have the problem of leading to node proliferation where mesh lines needed to provide a fine discretization in one part of the analysis domain can spill over into areas where a fine discretization is not necessary. A solution to this problem is to allow the mesh lines to terminate inside the area of fine discretization as in the finite box method [4.5]. The value of the approximating

Spatial Discretization

function at every missing node in the finite box technique is obtained by interpolating the corresponding neighbouring values. An even better solution would probably be to use triangular grids since that would provide the added benefit of being able to discretize irregular boundaries. The problem with these irregular discretization techniques is that they require a much larger amount of book keeping and cpu time in their formulation especially when dealing with three-dimensional space as is required in the substrate coupling problem.

To see how spatial discretization is employed in three dimensions, let us apply the box integration technique to (4.15) [4.1]-[4.5]. From Gauss' law (4.18),

$$\nabla \bullet E = k \qquad (4.21)$$

where $k = \rho/\varepsilon$.

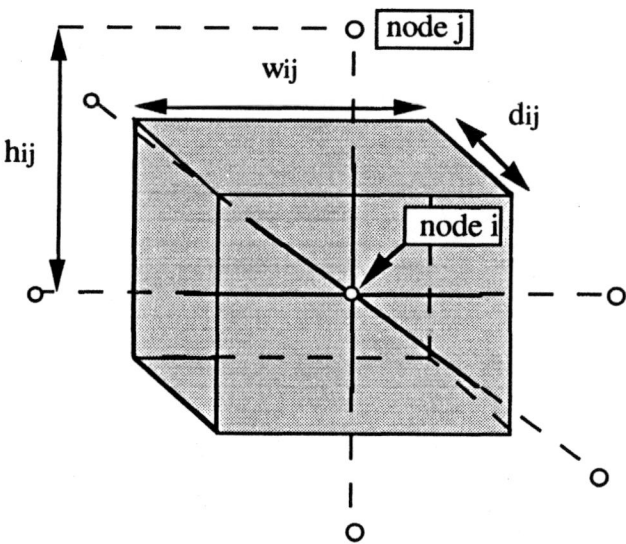

FIGURE 4.1 A control volume in the box integration technique.

Integrating $\nabla \bullet E$ over a volume Ω_i surrounding node i as shown in Figure 4.1,

$$\int_{\Omega_i} \nabla \cdot E \, d\Omega = \int_{\Omega_i} k \, d\Omega. \qquad (4.22)$$

From the divergence theorem

$$\int_{S_i} E \, dS = \int_{\Omega_i} k \, d\Omega \qquad (4.23)$$

where S_i is the surface area of the cube shown in Figure 4.1. Assuming k to be a constant in the volume Ω_i, the integral on the left hand side of (4.23) can be approximated as,

$$\sum_j E_{ij} \cdot S_{ij} = \sum_j E_{ij} \cdot w_{ij} d_{ij} = k \cdot \Omega_i \qquad (4.24)$$

and since $E_{ij}=(\Psi_i-\Psi_j)/h_{ij}$ (4.24) reduces to,

$$\sum_j \frac{w_{ij} d_{ij}}{h_{ij}} \cdot (\Psi_i - \Psi_j) = k \cdot \Omega_i. \qquad (4.25)$$

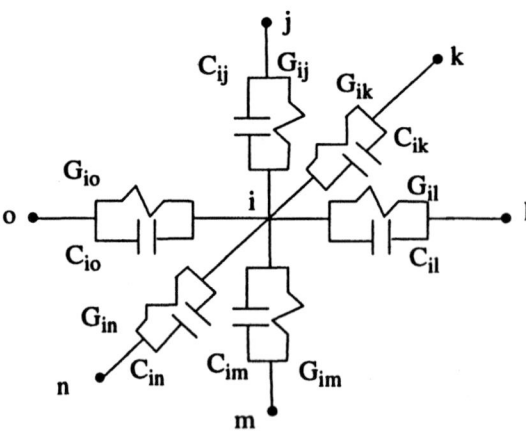

FIGURE 4.2 Resistances and capacitances around a node in substrate mesh.

Hence,

Boundary Conditions

$$\nabla \cdot E = k = \frac{1}{\Omega_i}\left(\sum_j \frac{w_{ij}d_{ij}}{h_{ij}}(\Psi_i - \Psi_j)\right) \quad (4.26)$$

and applying (4.26) to (4.15) results in

$$\sum_j \left[G_{ij}(\Psi_i - \Psi_j) + C_{ij}\left(\frac{\partial \Psi_i}{\partial t} - \frac{\partial \Psi_j}{\partial t}\right)\right] = 0 \quad (4.27)$$

where $G_{ij} = w_{ij}d_{ij}/\rho'h_{ij}$ and $C_{ij} = \varepsilon w_{ij}d_{ij}/h_{ij}$ as modelled with lumped circuit elements in Figure 4.2. Note that this model does not include any of the nonlinear parasitic capacitances associated with the device junctions as this is taken into account in the device model employed in the circuit simulator. It does however take into account field oxide capacitance and depletion capacitance of biased wells. Although depletion capacitance is actually nonlinear it is assumed that any noise voltage on the well bias is small signal and does not affect its capacitance value.

4.3 Boundary Conditions

Using some simplifying assumptions therefore, the substrate outside active areas can be treated as a linear 3D mesh where each mesh edge is a parallel combination of a resistor and a capacitor.

The active areas (contacts and devices) are treated as Dirichlet (fixed) boundaries for voltages,

$$\Psi = V_{appl} \quad (4.28)$$

and are connected either to the substrate/well nodes (for contacts) or the body nodes (for devices) in the equivalent schematic representation of the circuit (Figure 4.3). These Dirichlet boundaries will henceforth also be referred to as ports of the mesh.

Implicit in any discretization procedure is that all important physical phenomena are occuring within the region discretized; thus the surrounding environment can be ignored in the sense that no current is flowing and the electric potential is not changing significantly in these non discretized regions. Consequently at the edge surfaces (boundaries) of the substrate and the field oxide we impose Neumann (reflective) boundary conditions for both voltages and currents,

Simplified Substrate Modeling and Rapid Simulation

$$\frac{\partial \Psi}{\partial n} = \frac{\partial J}{\partial n} = 0 \qquad (4.29)$$

where n denotes the direction normal to the boundary.

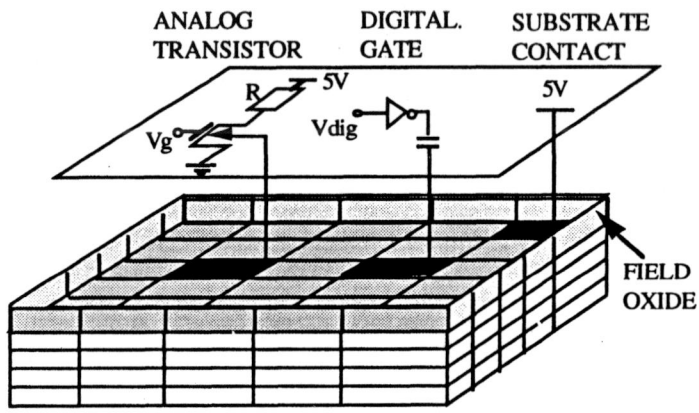

FIGURE 4.3 Substate mesh connected to the electrical circuit.

4.4 Solution Methods

Once the substrate is discretized into a 3D mesh it is necessary to simulate it in conjunction with the electrical circuit. Using conventional circuit simulation techniques [4.7], this system is solved by linearizing the circuit using the Newton-Raphson technique and employing a time integration technique to discretize the time derivatives of all the system state variables in order to formulate a dc admittance matrix that is solved iteratively at every time point in the analysis run using Gaussian elimination. The resulting matrix equation in modified nodal analysis form is

$$\begin{bmatrix} Y_{ckt} & \dfrac{\partial I_{ckt}}{\partial V_{mesh}} \\ \dfrac{\partial I_{mesh}}{\partial V_{ckt}} & Y_{mesh} \end{bmatrix} \cdot \begin{bmatrix} V_{ckt} \\ V_{mesh} \end{bmatrix} = \begin{bmatrix} I_{ckt} \\ I_{mesh} \end{bmatrix} \qquad (4.30)$$

Solution Methods

where Y_{ckt} and Y_{mesh} are the admittance matrices of the electrical circuit and mesh respectively and the terms $\partial I_{ckt}/\partial V_{mesh}$ and $\partial I_{mesh}/\partial V_{ckt}$ represent the influence of the mesh parameters on the circuit and the circuit parameters on the mesh respectively. The main problem with the matrix of (4.30) is that the submatrix Y_{mesh} is much larger than Y_{ckt} for a typical circuit. Moreover, since the time-discretized companion models of the energy storage elements in the circuit and mesh depend on the time-step taken at a particular instant the left hand side of (4.30) has to reevaluated at every successive time-point (SPICE-like simulators employ automatic variable time step control). Consequently, the computational effort required to solve this system of equations causes this solution procedure to be orders of magnitude slower than circuit simulation excluding substrate coupling effects.

In order to reduce the complexity of the problem, it is necessary to reduce the size of the matrix to be solved. It is easily seen that a majority of the nodes in the mesh do not directly interact with the circuit, i.e., entries in the submatrices $\partial I_{ckt}/\partial V_{mesh}$ and $\partial I_{mesh}/\partial V_{ckt}$ in (4.30) are non zero only at those nodes that physically connect the mesh to the circuit. Consequently, the mesh can be compacted into a much smaller admittance macromodel Y_{macro} which contains just adequate information to interface it to the external world. The macromodel is often smaller in size than the circuit being solved although it does decrease the sparsity of the overall matrix.

Since we are interested in analysis in the time-domain, it is necessary to develop a macromodel of the mesh that is valid across the transient analysis range. If a time integration algorithm is employed to discretize (in time) the capacitor voltages, there are two available choices- employ a fixed time-step or a variable time step. With the former methodology, the resulting macromodel is constant for the entire simulation range and needs to be computed just once. However, the time step of the whole system would be limited by the time step used by the mesh causing the simulation to slow down considerably. Using the latter methodology, it is possible to escape the time step limitation, but the macromodel becomes time-step dependent and requires to be recomputed at every time step.

Alternately, the admittance macromodel can be computed in the frequency domain as a set of transfer functions between the various ports in the mesh. Since frequency dependent elements can be expressed as a linear function of the complex frequency, s, each admittance parameter, y_{ij} in the macromodel (i and j correspond to its position in the matrix) can be expressed as a ratio of polynomials in s,

$$y_{ij}(s) = \frac{Q(s)}{P(s)}. \tag{4.31}$$

Simplified Substrate Modeling and Rapid Simulation

Simplified Substrate Modeling and Rapid Simulation

The roots of the denominator polynomial

$$P(s) = (s-p_1)(s-p_2)\ldots(s-p_n) = 0 \tag{4.32}$$

are the poles (or natural frequencies) of the circuit. An easy way to represent $y_{ij}(s)$ is in terms of its partial fraction expansion

$$y_{ij}(s) = \sum_{l=1}^{n} \frac{k_l}{s-p_l} \tag{4.33}$$

where k_l is the residue that corresponds to pole p_l. Using the inverse Laplace transform (4.33) could be converted into the time domain and simulated along with the circuit submatrix, Y_{ckt} to provide the system response. The biggest problem in this procedure is determining the poles of the system. Let the (modified) nodal analysis representation of the mesh equations be

$$Y_{mesh}(s) V_{mesh}(s) = I_{mesh}(s). \tag{4.34}$$

From Cramer's rule it is known that the response at any node j in the mesh is given by

$$V_j = \frac{det(T)}{det(Y_{mesh})} \tag{4.35}$$

where

$$T = \begin{bmatrix} y_{11} & \cdots & y_{1,j-1} & I_1 & y_{1,j+1} & \cdots & y_{1,n} \\ y_{21} & \cdots & y_{2,j-1} & I_2 & y_{2,j+1} & \cdots & y_{2,n} \\ \cdots & \cdots & \cdots & \cdots & \cdots & \cdots & \cdots \\ y_{n1} & \cdots & y_{n,j-1} & I_n & y_{n,j-1} & \cdots & y_{nn} \end{bmatrix} \tag{4.36}$$

with the column vector $I(s)$ replacing the jth column in $Y_{mesh}(s)$. It is apparent from (4.35) that the roots of $|Y_{mesh}(s)|$ are the poles of the circuit transfer function. So for a value $s=p_i$ where p_i is a circuit pole, $|Y_{mesh}(s)| = 0$ and therefore $Y_{mesh}(s = p_i)$ is singular. As in typical pole-zero analysis simulators, Muller's algorithm can be used to iteratively search for points in the s-plane where $Y_{mesh}(s)$ is singular. Starting with 3 points in the s-plane to evaluate $|Y_{mesh}(s)|$ an interpolating polynomial is formed to search for roots of $|Y_{mesh}(s)|$. The search is continued iteratively until $|Y_{mesh}(s)|$ is sufficiently small. To determine if $Y_{mesh}(s)$ is singular at a point an LU-factorization is attempted. If the factorization fails, it is assumed that the point represents a pole [4.9]. For a particular admittance parameter the zeroes of the transfer function (roots of $Q(s)$ in (4.31)) can also be determined using Muller's algorithm on the matrix $T(s)$ (of (4.36)). Once the poles and zeroes are determined, the residues of (4.33) are obtained trivially. Needless to say, determining the poles of a system is a fairly tedious and inefficient process, especially since some of them make an insignificant contribution to circuit performance. What we seek instead is an efficient means to obtain the few "dominant poles" that adequately characterize circuit behaviour.

4.5 Asymptotic Waveform Evaluation (AWE)

Asymptotic Waveform Evaluation (AWE) is a technique to approximate the frequency or time domain responses of large linear circuits in terms of a few poles and residues. This low order approximation forms a reduced-order approximation for the high order input circuit.

To begin, let us define the time moments of a function. The time domain moments of a signal $f(t)$ are identical to the Taylor series coefficients about $s=0$ of the waveform's Laplace transform, $F(s)$ [4.15]:

$$F(s) = \int_0^\infty f(t) e^{-st} dt = \int_0^\infty f(t) \left[1 - st + \frac{1}{2} s^2 t^2 - \dots \right] dt \qquad (4.37)$$

or

$$F(s) = \int_0^\infty f(t) dt - s \int_0^\infty t f(t) dt + \frac{1}{2} s^2 \int_0^\infty t^2 f(t) dt + \dots = m_0 + m_1 s + m_2 s^2 + \dots \qquad (4.38)$$

where

$$m_i = \frac{(-1)^i}{i!} \cdot \int_0^\infty t^i f(t) dt \qquad (4.39)$$

is the ith moment of $f(t)$, or the ith coefficient of the Taylor series expansion of $F(s)$ about $s=0$.

Simplified Substrate Modeling and Rapid Simulation

The AWE algorithm consists of two main parts- moment computation and moment matching. Moment computation involves computing moments of a circuit function in an efficient manner. Moment matching matches the moments of the circuit function with those of a reduced order model to fix the parameters of the reduced order model.

In general, a lumped linear time-invariant system can be described by a first-order matrix differential equation

$$Tx(t) + W\frac{dx}{dt} = b(t) \qquad (4.40)$$

$$y(t) = Cx(t) + d(t) \qquad (4.41)$$

irrespective of the way the circuit equations are formulated (e.g., modified nodal analysis [4.8], sparse tableau [4.10], etc.). The vector $x(t)$ contains the fundamental circuit variables of interest; these could be currents, voltages, charges, etc., depending on the formulation method while vector $y(t)$ contains the output (secondary) variables of interest. The matrix T represents the contributions of memoryless elements (such as resistances), the matrix W contains contributions from memory elements (capacitances and inductances) and the matrix C consists of the constitutive relations of the branch elements. The right hand side vectors $b(t)$ and $d(t)$ describe the influence of independent sources. Taking the Laplace transform of (4.40) and (4.41) and assuming zero initial conditions yields

$$(T + sW)X(s) = B(s) \qquad (4.42)$$

$$Y(s) = CX(s) + D(s) \qquad (4.43)$$

Equations (4.42) and (4.43) can be used to determine the impulse response of the circuit $h(t)$ (the output vector $y(t)$ when the input vector consists of impulses). If we assume that the Laplace transform of $x(t)$ has a Taylor series expansion about $s=0$, and that the input sources are impulses, we have

$$X(s) = X_0 + sX_1 + s^2 X_2 + \ldots \qquad (4.44)$$

and

$$B(s) = B_0 \qquad (4.45)$$

$$D(s) = D_0 \qquad (4.46)$$

where both (4.45) and (4.46) are independent of s. Substituting (4.44) and (4.45) in (4.42) yields

$$(T + sW)\left(X_0 + sX_1 + s^2X_2 + ...\right) = B_0 .\qquad(4.47)$$

Equating like powers of s in (4.47) we get

$$TX_0 = B_0 \text{ and} \qquad(4.48)$$

$$TX_k = -WX_{k-1} \text{ for } k > 0 \qquad(4.49)$$

Equations (4.48) and (4.49) can be solved recursively to determine the moments of $x(t)$. First (4.48) is solved for X_0. Note that this is equivalent to performing a steady-state solution of (4.40) with the input sources set to their DC values B_0. After X_0 has been detemined, it can be used in (4.49) to compute the right-hand side vector $-WX_0$; this allows us to solve for X_1. The procedure is continued recursively until $2q$ moments are found. Note that this recursive procedure can be done easily by solving the circuit (Equation (4.40)) with all the input sources set to zero (i.e, voltage sources shorted and current sources opened) and every capacitor (inductor) replaced by a voltage (current) source of value equal to the negative of the product of the previously computed moment and the capacitance (inductance). To determine the moments of the output vector $y(t)$ it is necessary to compute D_0. This can be done by solving for the output variables in the circuit with all state variables set to zero (i.e, capacitors replaced by zero-valued voltage sources and inductors by zero-valued current sources). It can be seen from (4.41) that the resulting values of the output variables will be equal to D_0. The moment computations can typically be done very efficiently since to solve the circuit only 1 LU factorization needs to be done. The same LU factors can be employed in the recursive procedure since only the right hand side changes in every iteration. An additional LU factorization however, is needed to compute the matrix D_0 since in this computation all the state variables are set to zero.

Unfortunately, moment computation of the substrate mesh is slightly more complicated than the general approach described above. The main problem tends to be that in the substrate mesh, in the well depletion regions and field oxide regions, there exist large capacitance only sub-meshes, consisting of both cutsets and loops of capacitors. Using the general approach to solve for the moments of this mesh will require special techniques [4.11] to avoid the problems introduced when capacitors are replaced by voltage and current sources in the recursive procedure outlined above. This can be quite a cumbersome task when there are several such loops or cutsets that first have to be identified and then modified to allow for a circuit solution. A better solution is to solve separately for the voltages in the capacitance only sub-mesh and the resistance/capacitance submeshes [4.4]. This solution procedure is outlined in Section 4.6. Another problem with macromodeling the substrate mesh is that the matrix that needs to be inverted tends to be extremely large and using direct solution methods such as LU factorization on it is prohibitive in terms of both cpu time and memory even if reordering and sparse matrix techniques are utilized. Section 4.9 outlines the iterative Incompete Choleski Conjugate Gradient (ICCG) method which works very efficiently for such matrices.

Once the $2q$ moments for the impulse response $H(s)$ have been determined

$$H(s) = m_0 + m_1 s + m_2 s^2 + ... \qquad (4.50)$$

the AWE algorithm matches these moments to a reduced order model, $\hat{H}(s)$ by using Padé approximation. The Padé approximant $\hat{H}(s)$ is a rational function

$$\hat{H}(s) = \frac{\beta_{q-1} s^{q-1} + ... + \beta_1 s + \beta_0}{\alpha_q s^q + ... + \alpha_1 s + 1}. \qquad (4.51)$$

The unknown coefficients $\alpha_i, i \in [1, q]$ and $\beta_i, i \in [0, q-1]$ are constrained so that the Maclaurin expansion of $\hat{H}(s)$ agrees with that of the actual system function as far as possible. Equating (4.50) and (4.51) yields

$$m_0 + m_1 s + m_2 s^2 + ... = \frac{\beta_{q-1} s^{q-1} + ... + \beta_1 s + \beta_0}{\alpha_q s^q + ... + \alpha_1 s + 1}. \qquad (4.52)$$

Cross-multiplying we find

$$(\alpha_q s^q + \alpha_{q-1} s^{q-1} + ... + \alpha_1 s + 1)\left(m_0 + m_1 s + m_2 s^2 + ...\right)$$
$$= \beta_{q-1} s^{q-1} + \beta_{q-2} s^{q-2} + ... + \beta_1 s + \beta_0. \qquad (4.53)$$

Matching the coefficients of $s^0, s^1, s^2, ..., s^{q-1}$ yields
$$\beta_0 = m_0$$
$$\beta_1 = m_0 \alpha_1 + m_1$$
$$...$$
$$\beta_{q-1} = m_0 \alpha_{q-1} + m_1 \alpha_{q-2} + ... + m_{q-1}$$
(4.54)

Asymptotic Waveform Evaluation (AWE)

Similarly, matching the coefficients of $s^q, s^{q+1}, \ldots, s^{2q-1}$ yields

$$0 = m_0\alpha_q + m_1\alpha_{q-1} + \ldots + m_{q-1}\alpha_1 + m_q$$
$$0 = m_1\alpha_q + m_2\alpha_{q-1} + \ldots + m_q\alpha_1 + m_{q+1}$$
$$\ldots$$
$$0 = m_{q-1}\alpha_q + m_q\alpha_{q-1} + \ldots + m_{2q-2}\alpha_1 + m_{2q-1}$$
(4.55)

which in matrix form is written as

$$\begin{bmatrix} m_0 & m_1 & \cdots & m_{q-1} \\ m_1 & m_2 & \cdots & m_q \\ \cdots & \cdots & \cdots & \cdots \\ m_{q-1} & m_q & \cdots & m_{2q-2} \end{bmatrix} \begin{bmatrix} \alpha_q \\ \alpha_{q-1} \\ \cdots \\ \alpha_1 \end{bmatrix} = - \begin{bmatrix} m_q \\ m_{q+1} \\ \cdots \\ m_{2q-1} \end{bmatrix}.$$
(4.56)

Once the α_i's of (4.56) are solved for, the β_i's can be solved for using (4.54) after which the denominator and numerator of (4.51) can be factored to obtain the poles, $\hat{p}_i$ and zeroes, $\hat{z}_i$ of the approximant, $\hat{H}(s)$ respectively. However, since we are more interested in the pole-residue format of the approximant, we can rewrite $\hat{H}(s)$ as

$$\hat{H}(s) = \sum_{i=1}^{q} \frac{k_i}{s - p_i}.$$
(4.57)

Expanding each of the partial fraction terms as a series in s yields,

$$\frac{k_i}{s - p_i} = -\frac{k_i}{p_i}\left(1 + \frac{s}{p_i} + \frac{s^2}{p_i^2} + \frac{s^3}{p_i^3} + \ldots\right).$$
(4.58)

Substituting (4.58) in (4.57) and replacing it in the right hand side of (4.52) gives

$$-\left(\frac{k_1}{p_1} + \frac{k_2}{p_2} + \ldots + \frac{k_q}{p_q}\right) = m_0$$

$$-\left(\frac{k_1}{p_1^2} + \frac{k_2}{p_2^2} + \ldots + \frac{k_q}{p_q^2}\right) = m_1 .$$
(4.59)

$$\vdots$$

Simplified Substrate Modeling and Rapid Simulation

Simplified Substrate Modeling and Rapid Simulation

Thus, the q unknown residues can be solved for using the first q equations in (4.59) as

$$\begin{bmatrix} 1 & 1 & \cdots & 1 \\ \dfrac{1}{p_1} & \dfrac{1}{p_2} & \cdots & \dfrac{1}{p_q} \\ \cdots & \cdots & \cdots & \cdots \\ \dfrac{1}{p_1^{q-1}} & \dfrac{1}{p_2^{q-1}} & \cdots & \dfrac{1}{p_q^{q-1}} \end{bmatrix} \begin{bmatrix} k_1 \\ k_2 \\ \cdots \\ k_q \end{bmatrix} = \begin{bmatrix} m_0 \\ m_1 \\ \cdots \\ m_{q-1} \end{bmatrix}. \tag{4.60}$$

Although this procedure appears to be a formidable task, typically q=4 is is the range of interest. The poles p_i and their residues k_i which result from the above are approximate dominant poles and corresponding residues for the circuit. In practice we start with a one pole approximation and proceed to higher orders until we are satisfied that no more useful information can be obtained.

4.6 Substrate AWE Macromodels

As discussed in Section 4.5, the general AWE approach does not work easily for the substrate mesh as it is difficult to find independent current loops since there are several capacitance-only loops and capacitance-only cutsets. In order to accomodate this it is necessary to redefine the AWE algorithm based upon a nodal analysis approach, since it is more suitable in solving (4.27) [4.4].

To begin, we classify all possible nodes in the substrate mesh using the following taxonomy:

cc - Nodes connected only to capacitances.

cg - Nodes connected to both resistances and capacitances.

gg - Nodes connected to resistances only.

By ordering nodes using the classification above, the following matrix equation can be obtained for the substrate mesh

Substrate AWE Macromodels

$$\begin{bmatrix} C_{cc-cc} & C_{cc-cg} & 0 \\ C_{cc-cg} & C_{cg-cg} & 0 \\ 0 & 0 & 0 \end{bmatrix} \begin{bmatrix} \dot{V}_{cc} \\ \dot{V}_{cg} \\ \dot{V}_{gg} \end{bmatrix} = - \begin{bmatrix} 0 & 0 & 0 \\ 0 & G_{cg-cg} & G_{cg-gg} \\ 0 & G_{cg-gg} & G_{gg-gg} \end{bmatrix} \begin{bmatrix} V_{cc} \\ V_{cg} \\ V_{gg} \end{bmatrix}$$

$$+ \begin{bmatrix} C_{cc-cc} & C_{cc-cg} & 0 \\ C_{cc-cg} & C_{cg-cg} & 0 \\ 0 & 0 & 0 \end{bmatrix} \begin{bmatrix} \dot{E}_{cc} \\ \dot{E}_{cg} \\ \dot{E}_{gg} \end{bmatrix} + \begin{bmatrix} 0 & 0 & 0 \\ 0 & G'_{cg-cg} & G'_{cg-gg} \\ 0 & G'_{cg-gg} & G'_{gg-gg} \end{bmatrix} \begin{bmatrix} E_{cc} \\ E_{cg} \\ E_{gg} \end{bmatrix}$$ (4.61)

where V_{xy} is the node voltage vector of nodes in class xy and E_{xy} is the excitation voltage vector applied to the nodes in class xy. G_{ab-xy} (C_{ab-xy}) consists of all resistances (capacitances) connected to one node in class ab and another in class xy, while G'_{ab-xy} (C'_{ab-xy}) consists of those resistances (capacitances) in G_{ab-xy} (C_{ab-xy}) that have a connection to an excitation source. In symbolic form (4.61) can be written as

$$C\dot{V} = -GV + B_1\dot{E} + B_0 E.$$ (4.62)

Due to the fact that not all node voltages are independent, both the G and C matrices are singular. To obtain the node voltage moments we see from (4.49) that $G^{-1}C$ must be calculated which of course is not possible since G is singular. The key idea in the following algorithm is that the relationship $Cm_{i-1} = -Gm_i$, where m_i is the ith moment, must hold even if C and G is singular. For simplicity, step inputs will be used in the algorithm. Moments of the impulse response are trivially obtained from those of the step response since $H(s) = sU(s)$, $U(s)$ being the step response of the mesh. If the step excitation vectors E_{cc} and E_{cg} are not zero, an impulse charging current flows into capacitors at $t=0$. To avoid this situation, the initial condition is taken at $t=0_+$ just after the impulse current flow goes to zero.

To compute the node voltage moments, it is necessary to first determine the initial state node voltages (i.e., at time $t = 0_+$). In the initial state it is assumed that charge has already been transferred by the impulse displacement current while the charge transfer by conduction current is still zero. Hence the voltage distribution is determined by the charge conservation law and the following set of equations is solved successively for the node voltages:

$$\begin{bmatrix} C_{cc-cc} & C_{cc-cg} \\ C_{cc-cg} & C_{cg-cg} \end{bmatrix} \begin{bmatrix} V_{cc}(0_+) \\ V_{cg}(0_+) \end{bmatrix} = \begin{bmatrix} E_{cc}(0_+) \\ E_{cg}(0_+) \end{bmatrix}$$

(4.63)

Simplified Substrate Modeling and Rapid Simulation

$$\begin{bmatrix} I & 0 \\ G_{cg-gg} & G_{gg-gg} \end{bmatrix} \begin{bmatrix} V_{cg}(0_+) \\ V_{gg}(0_+) \end{bmatrix} = \begin{bmatrix} V_{cg}(0_+) \\ E_{gg}(0_+) \end{bmatrix}. \quad (4.64)$$

Next the node voltages in the dc state ($t = \infty$) are determined. In steady-state, it is assumed that the displacement current has decayed and the voltage distribution is determined by KCL of the conduction current. The following set of equations is then solved successively

$$\begin{bmatrix} G_{cg-cg} & G_{cg-gg} \\ G_{cg-gg} & G_{gg-cg} \end{bmatrix} \begin{bmatrix} V_{cg}(\infty) \\ V_{gg}(\infty) \end{bmatrix} = \begin{bmatrix} E_{cc}(\infty) \\ E_{gg}(\infty) \end{bmatrix} \quad (4.65)$$

$$\begin{bmatrix} C_{cc-cc} & C_{cc-cg} \\ 0 & I \end{bmatrix} \begin{bmatrix} V_{cc}(\infty) \\ V_{cg}(\infty) \end{bmatrix} = \begin{bmatrix} E_{cc}(\infty) \\ E_{cg}(\infty) \end{bmatrix}. \quad (4.66)$$

The lowest order moment, m_{-1} for the step response of the node voltages can be computed as (See Appendix A)

$$m_{-1} = -(V(0_+) - V(\infty)). \quad (4.67)$$

For every higher order moment, m_i the following constraint must be satisfied

$$Cm_{i-1} = Gm_i. \quad (4.68)$$

The moments are computed by solving the following equations successively until i reaches the required number.

$$I_{cg} = \begin{bmatrix} C_{cc-cg} & C_{cg-cg} \end{bmatrix} \begin{bmatrix} m_{i-1, cc} \\ m_{i-1, cg} \end{bmatrix} \quad (4.69)$$

$$\begin{bmatrix} G_{cg-cg} & G_{cg-gg} \\ G_{cg-gg} & G_{gg-cg} \end{bmatrix} \begin{bmatrix} m_{i, cg} \\ m_{i, gg} \end{bmatrix} = -\begin{bmatrix} I_{cg} \\ 0 \end{bmatrix} \quad (4.70)$$

Substrate AWE Macromodels

$$\begin{bmatrix} C_{cc-cc} & C_{cc-cg} \\ 0 & I \end{bmatrix} \begin{bmatrix} m_{i,cc} \\ m_{i,cg} \end{bmatrix} = \begin{bmatrix} 0 \\ m_{i,cg} \end{bmatrix} \quad (4.71)$$

$$i = i + 1 \quad (4.72)$$

To determine the macromodel admittance parameters in $Y_{macro}(s)$, it is necessary to solve for the current moments at the ports of the mesh (Dirichlet boundaries at device/contact areas) as follows

$$\begin{aligned}
m_{i,cc(v),cg(v),gg(v)} &= \sum_{j=cg(v)} \left(\sum_{k=cg} (G_{cg-cg}[j][k] (m_{i,cg}[j] - m_{i,cg}[k]) \right. \\
&\quad + \left. \sum_{k=gg} G_{cg-gg}[j][k] (m_{i,cg}[j] - m_{i,gg}[k]) \right) \\
&\quad + \sum_{j=gg(v)} \left(\sum_{k=cg} G_{cg-gg}[j][k] (m_{i,gg}[j] - m_{i,cg}[k]) \right. \\
&\quad + \left. \sum_{k=gg} G_{gg-gg}[j][k] (m_{i,gg}[j] - m_{i,gg}[k]) \right) \\
&\quad + \sum_{j=cc(v)} \left(\sum_{k=cc} C_{cc-cc}[j][k] (m_{i-1,cc}[j] - m_{i-1,cc}[k]) \right. \\
&\quad + \left. \sum_{k=cg} C_{cc-cg}[j][k] (m_{i-1,cc}[j] - m_{i-(1,cg)}[k]) \right) \\
&\quad + \sum_{j=cg(v)} \left(\sum_{k=cc} C_{cc-cg}[j][k] (m_{i-1,cg}[j] - (m_{i-1,cc}[k])) \right. \\
&\quad + \left. \sum_{k=cg} C_{cg-cg}[j][k] (m_{i-1,cg}[j] - m_{i-1,cg}[k]) \right).
\end{aligned} \quad (4.73)$$

Although (4.73) appears to be complicated, it is not so if the notations are properly understood. Since each Dirichlet boundary may contain nodes from more than one category, the net current flowing through it is actually the sum of all the individual contributions. The subscripts $cc(v)$, $cg(v)$ and $gg(v)$ represent the nodes of the various categories that are located on the particular boundary in consideration. For each of these nodes, j, we sum up the currents flowing through each element connected to j and a surrounding node, k that is not of the same boundary. Note that since the current through a capacitor is the derivative of the voltage across it, the current moment due to it is equal to the product of the capacitance and the previous voltage moment. The voltage moment m_{-2} in (4.73) corresponds to the time derivative of voltage at $t=0_+$ and can be calculated as follows

Simplified Substrate Modeling and Rapid Simulation

$$I_{cg} = \begin{bmatrix} G_{gg-cg} & G_{cg-cg} \end{bmatrix} \begin{bmatrix} m_{-1,gg} \\ m_{-1,cg} \end{bmatrix} \quad (4.74)$$

$$\begin{bmatrix} C_{cc-cc} & C_{cc-cg} \\ C_{cc-cg} & C_{cg-cg} \end{bmatrix} \begin{bmatrix} m_{-2,cc} \\ m_{-2,cg} \end{bmatrix} = -\begin{bmatrix} 0 \\ I_{cg} \end{bmatrix} \quad (4.75)$$

Once the current moments to the step inputs are determined moments of the impulse response of the current are also obtained. The poles of this impulse response can be determined using (4.56) and then solving for the roots of the characteristic polynomial. Equation (4.60) is then used to obtain the correpsonding residues.

The substrate macromodel is computed as an $n \times n$ admittance matrix, $Y_{macro}(s)$ as in (4.76) (where n represents the number of ports) in which each of the n^2 parameters is represented by its AWE approximation.

$$\begin{bmatrix} y_{11}(s) & \cdots & y_{1n}(s) \\ \cdots & \cdots & \cdots \\ y_{n1}(s) & \cdots & y_{nn}(s) \end{bmatrix} \begin{bmatrix} v_1(s) \\ \cdots \\ v_n(s) \end{bmatrix} = \begin{bmatrix} i_1(s) \\ \cdots \\ i_n(s) \end{bmatrix} \quad (4.76)$$

The moments of the n^2 parameters are determined from the moments of the n port currents for n sets of voltage excitations. This is typically done by setting the jth port voltage to unity (unit step) and zeroing all others (i.e, shorting the ports). Then, the moments of the n port currents represent the moments of the admittance parameters of the jth column in $Y_{macro}(s)$. Repeating this pocedure with a unit voltage source at each of the other ports determines the moments of the remaining columns of parameters of $Y_{macro}(s)$. Thus for $i=1$ to n, $l=1$ to $2q$

$$m_l\{Y_{ij}(s)\} = m_l\{I_i(s)\} \quad (4.77)$$

$$V_j(s) = 1; V_k(s) = 0 \qquad k = 1 \text{ to } n; \; k \neq j \quad (4.78)$$

where $2q$ is the total number of moments.

4.7 Transient Simulation of AWE Macromodels

To simulate substrate coupling in any given circuit it is necessary to combine the substrate macromodel with the electrical (nonlinear) circuit. The contributions of the admittances of the macromodel must be stencilled into the admittance matrix of the electrical circuit to form a global Y-matrix at every time point in the simulation run. In a transient simulation, the values of all circuit variables must be determined at every time point from the beginning to the end of the analysis period. As the simulation proceeds, at the time point of interest, the values of all circuit voltages at every previous time point are known while those at the present time point are unknown. To see how the time domain representation of the AWE substrate macromodel is developed, we will first look at the general relationship between the frequency domain and time domain representation of an admittance macromodel. The frequency domain product

$$Y(s)V(s) = I(s) \qquad (4.79)$$

corresponds to a time domain convolution integral

$$i(t) = \int_0^t y(t-\tau)v(\tau)d\tau . \qquad (4.80)$$

In general, the convolution integral is computationally expensive especially as time progresses in the simulation run. However, in a circuit simulation context, the port voltages of the macromodel can be assumed to be linear functions of time between successive time points (Figure 4.4).

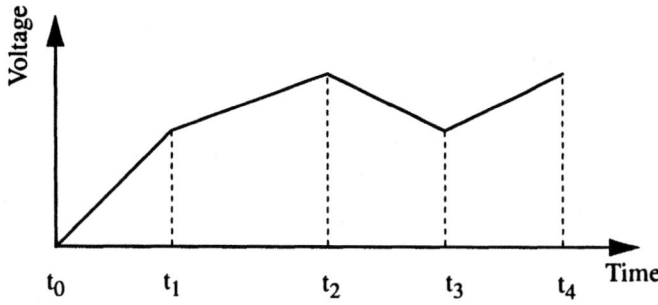

FIGURE 4.4 Piecewise linear port voltage waveform.

Subsequently, the piecewise linear waveform can be decomposed into a set of ramp functions, as shown in Figure 4.5, each begininng at a successive time point in the simulation run and each extending to infinity. After the initial ramp (i.e., the ramp function originating at the initial time point) where the slope of the ramp is equal to the slope of the port voltage waveform, the slope of every successive ramp is equal to the change in slope of the port voltage waveform at its time point of origin. Thus for any given time point the slopes of all but the final ramp function are known.

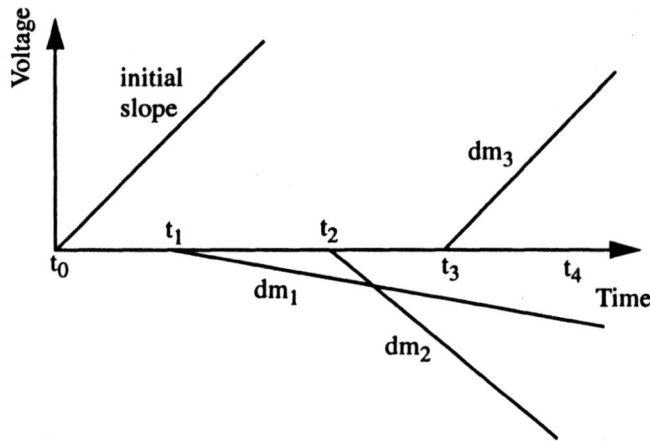

FIGURE 4.5 Decomposition of the port voltage waveform.

Since each term in the admittance matrix is in a partial fraction form, it is possible to find the frequency domain product of each ramp with each pole-residue pair and then find the inverse Laplace transform symbolically on a term-by-term basis. Assuming a single-pole admittance and an initial slope of m beginning at time t_0, the port current waveform due to the initial ramp is given by:

$$i(s) = \frac{m}{s^2} e^{-st_0} \cdot \frac{k}{s-p} . \qquad (4.81)$$

Applying the inverse Laplace transform

$$i(t) = m \left[-\frac{k}{p^2} + \frac{(-k)(t-t_0)}{p} + \frac{k}{p^2} e^{p(t-t_0)} \right] u(t-t_0) . \qquad (4.82)$$

Introducing the variables

$$d_1 = m\left[-\frac{k}{p^2} + \frac{k}{p}t_0\right] \quad (4.83)$$

$$e_1 = -m\frac{k}{p}t \quad (4.84)$$

$$f_1 = m\frac{k}{p^2} \quad (4.85)$$

the current $i(t_1)$ at time t_1 is given by

$$i(t_1) = d_1 + e_1 t_1 + f_1 e^{p(t_1 - t_0)}. \quad (4.86)$$

At every successive time point a new ramp with a slope equal to the change in slope at that time point is added. Therefore at time t_k the current $i(t_k)$ is computed as follows:

$$i(t_k) = d_k + e_k t_k + f_k e^{p(t_k - t_{k-1})} \quad (4.87)$$

where

$$d_k = d_{k-1} + dm_{k-1}\left(-\frac{k}{p^2} + \frac{kt_{k-1}}{p}\right) \quad (4.88)$$

$$e_k = e_{k-1} + dm_{k-1}\left(-\frac{k}{p}\right) \quad (4.89)$$

$$f_k = f_{k-1} e^{p(t_{k-1} - t_{k-2})} + dm_{k-1}\frac{k}{p^2}. \quad (4.90)$$

The unknown change in slope dm_{k-1} is given by

$$dm_{k-1} = \frac{v(t_k) - v(t_{k-1})}{t_k - t_{k-1}} - \frac{v(t_{k-1}) - v(t_{k-2})}{t_{k-1} - t_{k-2}}. \quad (4.91)$$

At time point t_k the voltage $v(t_k)$ is unknown. Equation (4.87) therefore consists of a known part and an unknown part. The coefficient of $v(t_k)$ (unknown part) is stencilled into the left hand side of the MNA equation $Y_{global}(t)v(t) = i(t)$ while the known part is stencilled into the right hand side. At each time point a constant number of operations is required to introduce an additional ramp into the stencil . This algorithm is fast and makes efficient use of computer memory [4.14].

4.8 Substrate DC Macromodels

The relaxation time of the substrate (outside of the depletion regions of p-n junctions) given by $\tau = \rho'\varepsilon$, with ρ' and ε as defined in Section 4.1, is of the order of $10^{-11}s$ (with $\rho' = 15\Omega\text{-}cm$ and $\varepsilon_0 = 11.9$). Consequently, the intrinsic substrate capacitances can be neglected for operating speeds of up to a few GHz and switching times of the order of 0.1 ns. Moreover, if the well capacitances, and capacitances to the substrate through the field oxide and die attach are modelled as lumped circuit elements, the substrate can be modelled as a purely resistive mesh [4.5]. In doing so additional boundary conditions need to be instituted. The well areas can be treated as Dirichilet boundaries for voltages and can be connected to an equivalent well capacitance externally. Since now the well capacitances are modelled external to the mesh, it is possible to model them as linear or nonlinear functions of voltage. Since the field oxide is no longer a part of the substrate model, a Neumann boundary is introduced at the top side of the substrate. The interconnect capacitance can be broken up into lumped equivalents and connected intermittently to the substrate. The die-attache capacitance can also be modelled as a lumped circuit element if a Dirichilet boundary condition is introduced at the backside of the die.

By moving the capacitive elements into the domain of the nonlinear simulator (in this case SPICE), we have reduced the substrate into a mesh which can be represented by a DC macromodel. While in the AWE macromodel several matrix solutions are required in computing the moments of the mesh and further computations in the moment matching algorithm for each admittance parameter, a dc admittance parameter is obtained using a single matrix solution of the mesh. Transient simulation of the dc macromodel is also simple- the admittance values of the parameters need to be introduced into the global admittance matrix at every time point. This could result in a significant reduction in cpu time especially with the moment computation algorithm of Section 4.6 and the matrix solution technique outlined in Section 4.9. Additionally, since the dc macromodel is essentially a network of resistors, it offers the advantage of portability as it can be used in conjunction with any nonlinear simulator and parasitic layout extractor. However, by discretizing further the continuous nature of substrate coupling, the accuracy of the resulting lumped models may be questionable. A solution to the problem may be to impose multiple boundary conditions on large well areas or the backside die-attache area at a larger computational expense. Since ultimately, the price to be paid is the cpu time complexity involved in the modeling/simulation approach, the model chosen should reflect this tradeoff. In the rest of this book however, we will limit our discussion to a resistive substrate.

4.9 Matrix Solution

The direct solution method for a matrix, A comprises decomposing the matrix into its lower diagonal and upper diagonal factors ($A = LU$) and applying a forward substitution and a back substitution to solve for the matrix equation. For the equation $Ax = b$, we would first solve via forward substitution, $Ly = b$ and then by back substitution, $Ux = y$. While this method has the advantage of reusability of the LU factors for different right hand side vectors, b, the computational effort involved in LU decomposition is of the order of $n^{1.1}$ to $n^{2.3}$ where n is the size of the matrix depending on its sparsity. To solve a 3D mesh where n is very large and the operation count is $O(n^2)$-$O(n^{2.3})$ the direct solution method is infeasible.

4.9.1 Iterative methods

An option to the direct solution method is the general class of iterative methods such as Gauss-Jacobi, Gauss-Seidel and Simultaneous Over-Relaxation (SOR) [4.12]. The general approach in these methods is to split the matrix, A into three parts:

$$A = L + D + U \qquad (4.92)$$

where D consists of the diagonal elements of A while L and U are strictly lower and upper triangular matrices respectively. The matrix equation $Ax = B$ can be rewritten as

$$(D + L + U) x = b \qquad (4.93)$$

or

$$Dx = -(L + U) x + b \qquad (4.94)$$

In the Gauss-Jacobi method, the iteration employed in determining the solution x is

$$Dx_{i+1} = -(L + U) x_i + b \ . \qquad (4.95)$$

Simplified Substrate Modeling and Rapid Simulation

Substituting $L+U=A-D$ in the right hand side of (4.95) gives

$$x_{i+1} = x_i - D^{-1}(Ax_i - b) \tag{4.96}$$

In the Gauss-Seidel method, the iteration is modified as

$$x_{i+1} = x_i - (L+D)^{-1}(Ax_i - b) . \tag{4.97}$$

It can be shown (see Appendix B) for the iterative methods that the error $e_i = x - x_i$ at the ith iteration converges according to the formula

$$e_i = \Lambda^i e_0 \tag{4.98}$$

where Λ is a diagonal matrix whose entries are the eigenvalues λ_j of the iteration matrix, M where $M = -D^{-1}(L+U)$ for Gauss-Jacobi and $-(L+D)^{-1}U$ for Gauss-Seidel [4.9]. Equation (4.98) implies that the iteration methods converge only if

$$\max_j |\lambda_j| < 1 \tag{4.99}$$

which is ensured if the matrix, A is diagonally dominant or positive definite. However, rate of convergence depends greatly on the values of the eigenvalues of M. Consequently, the components of the error vector, e that correspond to small eigenvalues fall off rapidly after a few iterations. After that the only components of e left correspond to eigenvalues that are just slightly less than 1 in modulus. Rate of convergence is worsened if there is large variation in component values in the matrix to be solved, as could be the case in a substrate with widely varying doping densities.

The SOR method tries to solve the convergence rate problem of the other two methods by using a parameter to control the size of the updates.

$$x_{i+1} = x_i - \omega(L+D)^{-1}(Ax_i - b) \tag{4.100}$$

For a slow but monotonic convergence, $\omega > 1$ (overrelaxation) would help converge faster by increasing the size of the updates. If the iterates execute damped oscillations about the solution, $\omega < 1$ (under-relaxation) would accelerate the convergence by reducing the size of the updates. Unfortunately, the optimum value of ω to use varies from case to case and depends on the eigenvalues of the iteration matrix which is dificult to compute.

Matrix Solution

4.9.2 Strongly Implicit Procedures (SIPs)

The idea in SIPs is to find "approximate" LU factors of A that are easy to compute and store. In other words, we find $\hat{L}$ and $\hat{U}$ such that

$$A = \hat{L}\hat{U} + R \tag{4.101}$$

where R is small compared to A. The iterative method is derived as follows:

$$Ax = b \tag{4.102}$$

$$(\hat{L}\hat{U} + R)x = b \tag{4.103}$$

$$\hat{L}\hat{U}x = -Rx + b \tag{4.104}$$

$$\hat{L}\hat{U}x_{i+1} = -Rx_i + b. \tag{4.105}$$

$$\hat{L}\hat{U}(x_{i+1} - x_i) = -Ax_i + b \tag{4.106}$$

This equation can be solved with forward and back substitutions with the approximate LU factors. Clearly if $LU = A$, the solution process would require just one iteration. In practice, the LU factors are computed by ignoring fill-ins when they occur in the LU decomposition process. This allows the SIP method to retain the memory advantage of iterative solvers. In practice, SIP methods are only marginally better than Gauss-Jacobi or Gauss-Seidel because to construct incomplete LU factors a lot of fill-ins are ignored leading to a poor approximation, requiring many iterations.

4.9.3 Gradient Descent Methods

This is a general class of matrix solvers that model the matrix solution problem as an optimization problem [4.13]. For a positive definite, symmetric matrix, A as is for a nodal admittance matrix of a passive circuit, defining a function $f(x)$ as

$$f(x) = \frac{1}{2}x^T A x - b^{Tx} \tag{4.107}$$

we see that the gradient of the function is

$$\nabla f(x) = \frac{1}{2}\left(A + A^T\right)x - b = Ax - b \tag{4.108}$$

Simplified Substrate Modeling and Rapid Simulation

for symmetric A. Therefore $f(x)$ has a unique minimum of 0 when $x = x^* = Ax - b$. To obtain the minimum of a function starting from an arbitrary starting point, line search algorithms are generally employed. This involves finding a direction and a step size to move to a new point close to the minimum. The simplest of these methods is the *steepest descent method*. Since the gradient of a function points in the direction of the maximum change in the value of that function, the negative of the gradient is chosen as the logical direction to move in to find the minimum. Once a direction is fixed, the algorithm finds the minimum of the function in that particular direction and selects it as the next solution update. The gradient of the function is once again computed at this new update and the procedure repeated until convergence is attained. The problem with the method is that once a minimum has been obtained in a fixed direction, the gradient of the function computed at this point will be orthogonal to the previous search direction (Otherwise the minimization is not complete in that direction). Consequently, the steepest descent method must execute a right angle turn at every iteration. This can consume a large amount of computational time if the function to be minimized has elongated contours and/or a bad initial guess is made.

A better strategy is the *conjugate gradient* method. The basic idea in this method is that we require the change in gradients of the function (from the present to the next update) to be orthogonal to the present direction since otherwise more movement in the present direction would be needed. Since $\nabla f = Ax - b$ with u as the present direction and v as the next direction we get

$$\delta(\nabla f) = [A(x+v) - b] - [Ax - b] = Av \qquad (4.109)$$

Also since ∇f is orthogonal to u we get

$$u^T(\nabla f) = 0 \Rightarrow u^T A v = 0. \qquad (4.110)$$

If this relationship holds, u and v are said to be *A conjugate* or simply *conjugate*. Letting the new search direction be a linear combination of the gradient and the old search vector

$$s_i = \nabla f + \beta s_{i-1} \qquad (4.111)$$

we impose the conjugacy requirement

$$s_i^T A s_{i-1} = (\nabla f + \beta s_{i-1})^T A s_{i-1} = 0 \qquad (4.112)$$

which can then be solved for β.

Matrix Solution

$$\beta = -\frac{(\nabla f)^T A s_{i-1}}{s_{i-1}^T A s_{i-1}}$$ (4.113)

Once the direction is fixed the new update x_{i+1} is taken to be

$$x_{i+1} = x_i + \alpha s_i$$ (4.114)

where α is chosen so as to minimize the function, $f(x)$ in the new direction s_i. Since

$$f(x_i + \alpha s_i) = \frac{1}{2}(x_i + \alpha s_i)^T A (x_i + \alpha s_i) - b^T (x_i + \alpha s_i)$$ (4.115)

or

$$f(x_i + \alpha s_i) = \left(\frac{1}{2} x_0^T A x_0 - b^T x_0\right) + \alpha s_i^T (A x_0 - b) + \frac{1}{2}\alpha^2 s_i^T A s$$ (4.116)

setting the derivative of $f(x)$ with respect to α to 0 gives

$$\frac{\partial}{\partial \alpha} f(x_0 + \alpha s_i) = \alpha s^T A s_i + s_i^T (A x_0 - b) = 0.$$ (4.117)

Solving for α yields

$$\alpha = -\frac{s_i^T (A x_0 - b)}{s_i^T A s_i}.$$ (4.118)

The nice feature of the conjugate gradient method is that for an $N \times N$ matrix it converges to the minimum within N steps. There are several cases in which convergence occurs even quicker. In particular if the matrix, A happens to have all eigenvalues equal the conjugate gradients converges in one step.

4.9.4 Incomplete Choleski Conjugate Gradient (ICCG) Method

J.A. Meijerink and H.A. van der Vorst combined the conjugate gradient method with strongly implicit procedures (SIPs) to develop the so-called ICCG method. The idea is to use approximate LU factors to precondition the matrix A, and then to use the matrix $(LU)^{-1}A$ in the conjugate gradient iteration. The advantage of this approach is that if the product of the incomplete LU factors $\hat{L}\hat{U}$ is a good approximation of A,

$$(\hat{L}\hat{U})^{-1} A \approx I, \quad (4.119)$$

the identity matrix which has all eigenvalues equal to 1 and the conjugate gradient method gives fast convergence for matrices with the latter property.

The term "Incomplete Choleski" in the name refers to a special type of LU decomposition, the Choleski decomposition, which only works for symmetric positive definite matrices. A matrix of this type can be decomposed as

$$A = LL^T. \quad (4.120)$$

The advantage of Choleski decomposition is that it requires only half the storage and operations of an asymmetric LU decomposition. It however, is not a requirement in the ICCG method and can be repaced by asymmetric incomplete LU decomposition. The ICCG method is as follows:

With an arbitrary initial approximation to x, x_0 determine the residual term, r_0

$$r_0 = b - A x_0 \quad (4.121)$$

and initial search direction p_0 as

$$p_0 = \left(LL^T\right)^{-1} r_0. \quad (4.122)$$

Then the following equations are computed recursively.

$$\alpha_i = \frac{r_i^T \left(LL^T\right)^{-1} r_i}{p_i^T A p_i} \quad (4.123)$$

$$x_{i+1} = x_i + \alpha_i p_i \quad (4.124)$$

Substituting for x_{i+1} using (4.124), the new residual $r_{i+1} = b - Ax_{i+1}$, is obtained as

$$r_{i+1} = r_i - \alpha_i A p_i. \tag{4.125}$$

The new direction p_{i+1} is computed as

$$p_{i+1} = \left(LL^T\right)^{-1} r_{i+1} + \beta_i p_i \tag{4.126}$$

where

$$\beta_i = \frac{r_{i+1}^T \left(LL^T\right)^{-1} r_{i+1}}{r_i^T \left(LL^T\right)^{-1} r_i}. \tag{4.127}$$

The ICCG method is guaranteed to converge to a solution in n iterations where n is the size of the matrix, A. Experience with ICCG shows that typical iteration counts are far fewer.

4.10 Results

To validate the macromodeling strategy, simulation results have been compared both to results from the device simulation program MEDICI [4.16], and also to results from measurements reported on an experimental chip [4.17].

4.10.1 Comparisons with Device Simulation

As seen in Chapter 3 the device simulation program models two-dimensional distributions of potential and carrier concentrations in a device to predict its electrical characteristics for any bias condition. It solves Poisson's equation and both the electron and hole continuity equations using numerical simulation techniques to analyze devices such as diodes, BJTs, MOSFETs etc. for dc, steady-state or transient operating conditions. Since mixed-mode ICs are generally fabricated in processes with either a heavily doped bulk with an epitaxial layer or a lightly doped substrate (without an epitaxial layer), circuits representative of both processes have been verified using the simulation tool. The doping profile of a 2 μm BiCMOS technology [4.3] is used for both simulations with MEDICI and the macromodel.

Simplified Substrate Modeling and Rapid Simulation

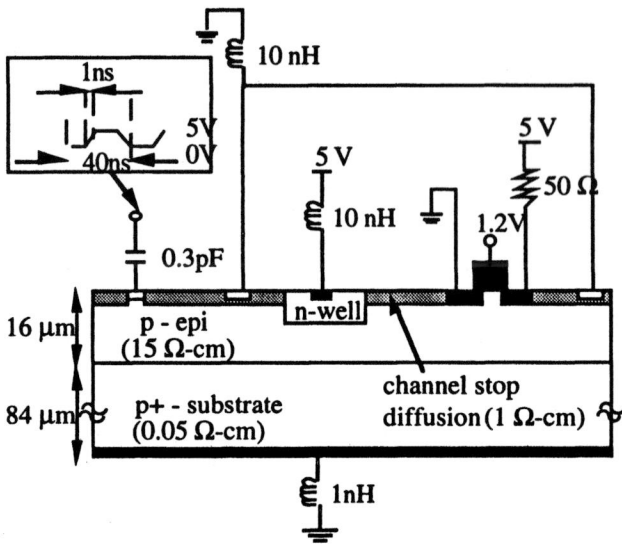

FIGURE 4.6 Circuit schematic/layout profile for simulations with a heavily doped bulk and a lightly doped epitaxial layer.

Figure 4.6 shows the experimental setup used to simulate substrate coupling in a heavily doped substrate (P+ bulk) with a lightly doped epitaxial layer (P- epi) [4.18],[4.2]. It consists of a diffused region equivalent to the drain of a switching transistor and a single NMOS transistor current source considered to be part of a sensitive analog circuit separated by a distance of 30 μm. Several established shielding techniques to reduce the coupling from the switching node to the sensitive transistor, including increased separation, an n-well diffusion, a p+ ring and a p+ contact strapping the backside of the substrate to ground potential have been tested with our macromodelling technique. Figures 4.7 and 4.8 compare the results obtained with the macromodeling technique to those obtained with the device simulation program and plots the drain noise voltage (peak-to-peak) and settling time behaviour of the sensitive transistor as functions of the shielding technique used. The separation between the switching and sensitive nodes is 30 μm unless otherwise specified. The substrate contact on the far right of Figure 4.6 is present in all the cases of Figure 4.7 and Figure 4.8. In case A there is no guard ring or backside contact between the switching and sensitive nodes. Case C shows the effect of simply increasing the separation between the two nodes to 200μm. In case D a backside contact is used to additionally bias the substrate. In case B an 8 μm wide n-well is placed midway between the

Results

switching and sensitive nodes while in case E an additional p+ contact is placed between the nodes.

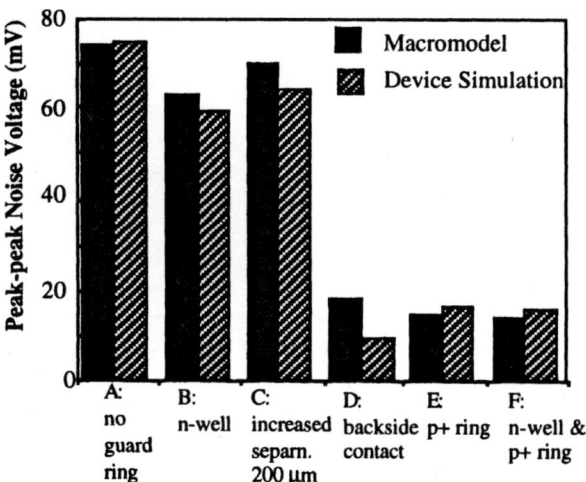

FIGURE 4.7 Effect of various shielding techniques on peak-peak noise voltage at the sensitive node in Figure 4.6.

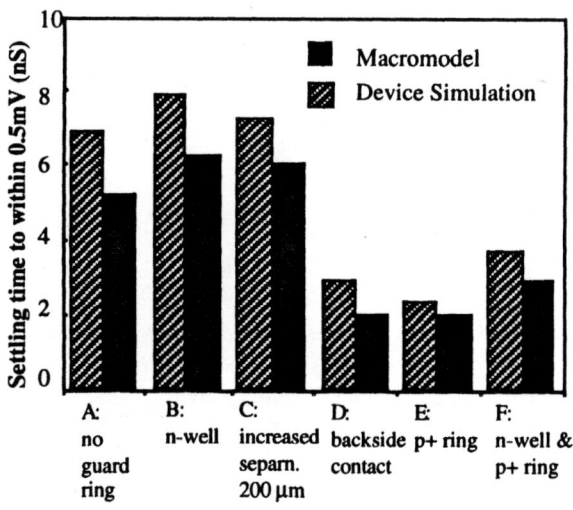

FIGURE 4.8 Effect of various shielding techniques on settling time of the noise voltage at the sensitive node in Figure 4.6.

Simplified Substrate Modeling and Rapid Simulation

Without any shielding, there is almost 75 mV of noise on the drain voltage of the sensitive transistor. Increased separation is not an effective means of reducing the noise because most of the coupling occurs through the heavily doped bulk which provides a low resistance path between the noisy and sensitive nodes independent of separation. The n-well diffusion placed midway between the noisy and sensitive nodes is also ineffective because most of the noise-coupling is due to majority carriers. On the other hand, a p+ ring and a backside contact are very effective since they provide low impedance paths from the power supply to the noisy bulk. Using both the n-well and p+ rings provides little additional improvement.

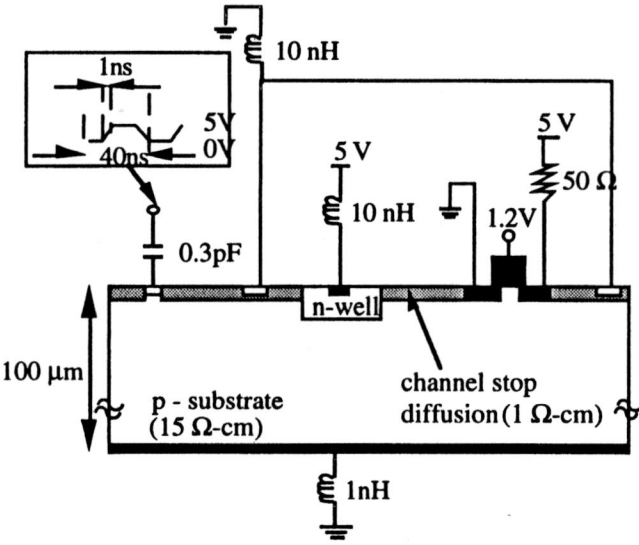

FIGURE 4.9 Circuit schematic/layout profile for simulations with a lightly doped substrate.

Figure 4.9 is the setup used for simulations with a uniformly lightly doped substrate (Bulk P- wafer) [4.2],[4.5] and is otherwise identical to Figure 4.6. Figure 4.10 shows the peak-peak voltage and Figure 4.11 the settling time behaviour of the resulting noise voltage waveforms using both the macromodelling technique and the device simulation program. As is evident in Figure 4.10, increased separation is a more effective means of shielding in lightly-doped substrates because the substrate resistance is more dependent on distance, owing to its higher resistivity. An n-well diffusion is again ineffective while a p+ ring provides a good shield from the substrate noise. A backside contact is not a very effective guarding configuration in a lightly-doped substrate because of the high resistance between the backside and the surface of the substrate.

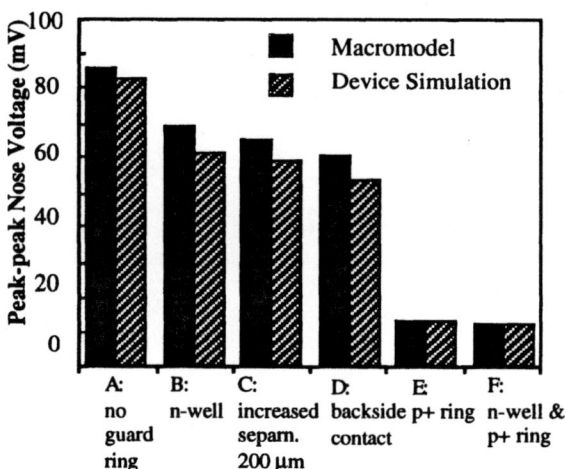

FIGURE 4.10 Effect of various shielding techniques on peak-peak noise voltage at the sensitive node in Figure 4.9.

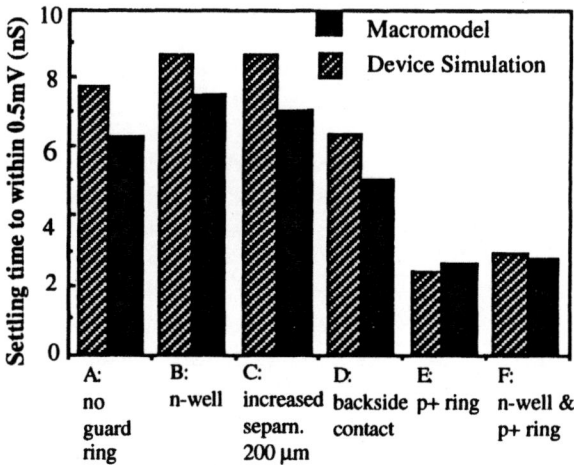

FIGURE 4.11 Effect of various shielding techniques on settling time of the noise voltage at the sensitive node in Figure 4.9.

Table 4.1 compares the cpu time requirements for the device simulation program and the AWE (2nd order) and DC macromodeling techniques. With the DC macromodeling technique, the n-well diffusions of Figure 4.6 and Figure 4.9 were extracted as ports and connected to lumped capacitances equivalent to the depletion capacitance of the n-well. Both the AWE and DC macromodeling techniques produced identical results for the examples shown. As exemplified in Table 4.1 , the main advantage in using the macromodeling techniques is the significantly shorter cpu times required in their analysis as compared to the device simulation program.

TABLE 4.1. Run-Time (on DECstation 5000) comparison between the device simulation program and the Macromodeling Techniques.

Number of mesh nodes	Device Simulation cpu time (s)	AWE Macromodel cpu time (s)	DC Macromodel cpu time (s)
2940	4375	60.6	5.5
3716	7192.8	92.3	7.2
6605	15882.2	202.2	20.3
8712	23732.2	294.3	28.3

4.10.2 Comparisons with Measured Results

The DC macromodeling technique has been used to simulate substrate coupling on the experimental chip reported in [4.17]. The 2 mm x 2 mm test chip realized in a 2 µm BiCMOS n-well process consists of transistors fabricated in a 15 µm lightly-doped epitaxial layer over a heavily-doped bulk. An on-chip ring oscillator drives a block of 12 CMOS inverters with each inverter output capacitively coupled to the substrate.The switching noise introduced into the substrate is measured by ten single transistor NMOS current sources distributed across the chip. The substrate is biased using a combination of several p+ contacts on the die surface. Seven of the current sources are shielded from the substrate noise using guard rings placed at varying distances (6 µm or 22 µm) from the sources and biased either with a dedicated package pin or connected to two large substrate contacts (one located at the chip center and one diffusion ring surrounding the chip).

The measured results [4.17] and simulated results are in good agreement as shown in Figure 4.12 although only approximate values have been used for bonding pad and chip-to-package capacitance and bond-wire/package pin inductances. When a guard ring is biased with a dedicated package pin (and supply) the noise voltage reduces by approximately 20% when the guard ring is placed 6 μm from the transistor. When the guard ring is connected to the large substrate contacts, the noisy power supply is brought in the vicinity of the sensitive transistor, causing an increase in the peak-peak noise voltage.

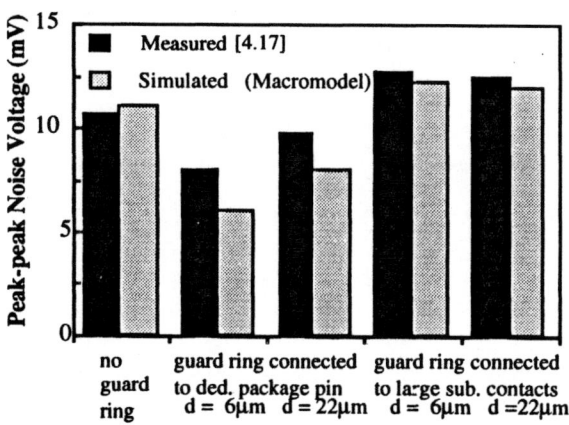

FIGURE 4.12 Effect of various guarding configurations on the drain noise voltage of an NMOS transistor on the test chip [4.17].

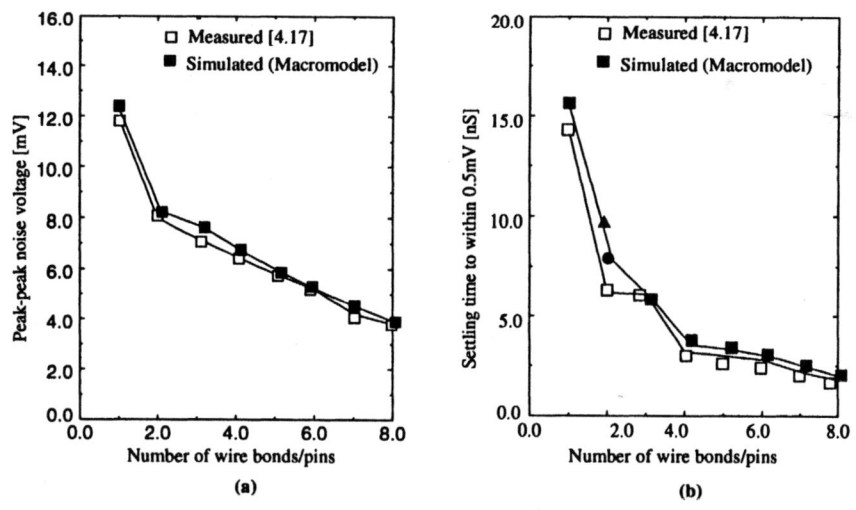

FIGURE 4.13 Effect of multiple substrate bias package pins /bond wires on (a) peak-peak noise voltage and (b) settling time of noise voltage.

Simplified Substrate Modeling and Rapid Simulation

Figure 4.13 shows the effect of reducing the power supply inductance on the noise voltage of a sensitive transistor. By increasing the number of package pins (10nH per bond wire package pin) used to bias the substrate power supply, noise voltage is reduced dramatically. Figure 4.13 compares the peak-peak and settling time behaviour of the measured noise voltage [4.17] with the simulated noise voltage.

4.11 Summary

This chapter introduced an alternative solution strategy to numerical simulation of the semiconductor device equations for the substrate coupling simulation problem. A simplified modeling strategy was developed and means to simulate it were discussed. In particular, Asymptotic Waveform Evaluation (AWE) was described and macromodeling the substrate using AWE was described. Also transient simulation of AWE macromodels alongwith the nonlinear electrical circuit was discussed. Further simplifications to the RC model and their implications were presented. Efficient matrix solution was identified as a critical part of the simulation strategy and different techniques for the same were reported and compared. The Incomplete Choleski Conjugate Gradient (ICCG) method was shown to be optimal for the substrate mesh and the algorithm was described. Simulations using the AWE macromodeling and DC macromodeling techniques were found to be both fast and accurate as compared to a device simulation program. These techniques were also compared to reported measurements on a test chip and found to be accurate.

REFERENCES

[4.1]   N.K. Verghese, S.S. Lee and D.J. Allstot, " A Unified Approach to Simulating Electrical and Thermal Substrate Coupling Interactions in ICs," *Proceedings of the International Conference on Computer-Aided Design*, pp. 422-426, 1993.

[4.2]   N.K. Verghese, *A Simulation Strategy for Substrate Coupling in Integrated Circuits*, Master's thesis, Carnegie Mellon University, 1993.

[4.3]   S. Masui, "Simulation of Substrate Coupling in Mixed-Signal MOS Circuit, " *Technical Digest of the VLSI Circuits Symposium*, pp. 42-43, June 1992.

[4.4] S. Kumashiro, *Transient Simulation of Passive and Active VLSI Devices Using Asymtotic Waveform Evaluation*, Ph.D. thesis, Carnegie Mellon University, 1992.

[4.5] B.R. Stanisic, N.K. Verghese, R.A. Rutenbar, L.R. Carley and D.J. Allstot, "Addressing Substrate Coupling in Mixed-Mode ICs: Simulation and Power Distribution Synthesis," *IEEE Journal of Solid-State Circuits*, vol. 29, no. 3, pp. 226-238, April 1994.

[4.6] M.S. Adler, "A Method for Terminating Mesh Lines in Finite Difference Formulation of the Semiconductor Device Equations," *Solid-State Electronics*, vol. 23, pp. 845-853, 1980.

[4.7] L.W. Nagel and D.O. Pederson, "Simulation Program with Integrated Circuit Emphasis [SPICE]," *Proceedings of the 16th Midwest Circuits Symposium on Circuit Theory*, 1973.

[4.8] C.W. Ho, A.E. Ruehli and P.A. Brennan, "The Modified Nodal Analysis Approach to Network Analysis," *IEEE Transactions on Circuit Theory*, vol. CAS-22, pp. 504-509, June 1975.

[4.9] R. A. Rohrer, *Circuit Simulation- Theory and Practice*, Class Notes, 1991.

[4.10] G.D. Hachtel, R.K. Brayton and F.G. Gustavson, "The Sparse Tableau Approach to Network Analysis and Design," *IEEE Transactions on Circuit Theory*, CT-18101-113, 1971.

[4.11] X. Huang, *Padé Approximation of Linearized Circuit Responses*, Ph.D. dissertation, Carnegie Mellon University, December 1990.

[4.12] E. Isaacson and H.B. Keller, *Analysis of numerical methods*, John Wiley & Sons, 1966.

[4.13] R. Fletcher, *Practical Methods of Optimization*, John Wiley & Sons, 1987.

[4.14] V. Raghavan, J.E. Bracken and R.A. Rohrer, "AWESpice: A general tool for the accurate and efficient simulation of interconnect problems," *Proceedings of the ACM/IEEE Design Automation Conference*, pp. 87-92, 1992.

[4.15] V. Raghavan, *An Integrated Methodology for Interconnect Analysis*, Ph. D dissertation, Carnegie Mellon University, December '92.

[4.16] *TMA MEDICI : Two Dimensional Device Simulation Program*, Version 1, Volume 1, Technology Modeling Associates, Inc., 1992.

[4.17] D.K. Su, M.J. Loinaz, S. Masui and B.A. Wooley, "Experimental Results and Modeling Techniques for Substrate Noise in Mixed-Signal Integrated Circuits," *IEEE Journal of Solid State Circuits*, vol. 28, no. 4, April 1993.

[4.18] N.K. Verghese, D.J. Allstot and S. Masui, "Rapid Simulation of Substrate Coupling Effects in Mixed-Mode ICs," *Proceedings of the Custom Integrated Circuits Conference*, pp. 18.3.1-18.3.4, May 1993.

CHAPTER 5   *Mesh Generation*

An important part of the simulation procedure is the generation of the mesh to be solved. From the discussion on box integration in Section 4.2 , we note that we have assumed the electric field intensity vector, E to be constant between any two adjacent nodes in the mesh. In other words, although the electric field varies nonlinearly as a function of distance, the spatial discretization method approximates it as a piecewise constant function. Obviously then, the discretization method is only as accurate as this approximation is and it becomes necessary to use fine grids to accurately approximate the electric field in regions where it is highly nonlinear. Wherever coarse grids are adequately accurate they should be used since the overall density of grids is a primary deterninant of the computational complexity involved in the solution procedure. However, since the field intensity cannot be determined before discretization, the density of grids needed is not known *a priori*. In general, there are two approaches to solving this problem.

5.1 Adaptive Mesh Refinement

In an adaptive refinement scenario one starts with a coarse mesh, solves the problem, evaluates the accuracy of the solution, adds nodes to the mesh in appropriate locations, resolves the problem and evaluates its accuracy, leading to further refinement until the answer is within acceptable accuracy limits. The procedure requires a way to determine where the solution accuracy is adequate and it is an open question as to what is the correct criterion to refine the grid. One possible method is to refine the mesh everywhere, solve it and determine the error between the new solution and the old solution. (Note that since the new mesh has extra nodes the old solution has to be interpolated to obtain solution values at these nodes) The error vector can then be utilized to further refine the mesh in areas where the error in the solution is large. Unfortunately, this is a prohibitively expensive procedure.

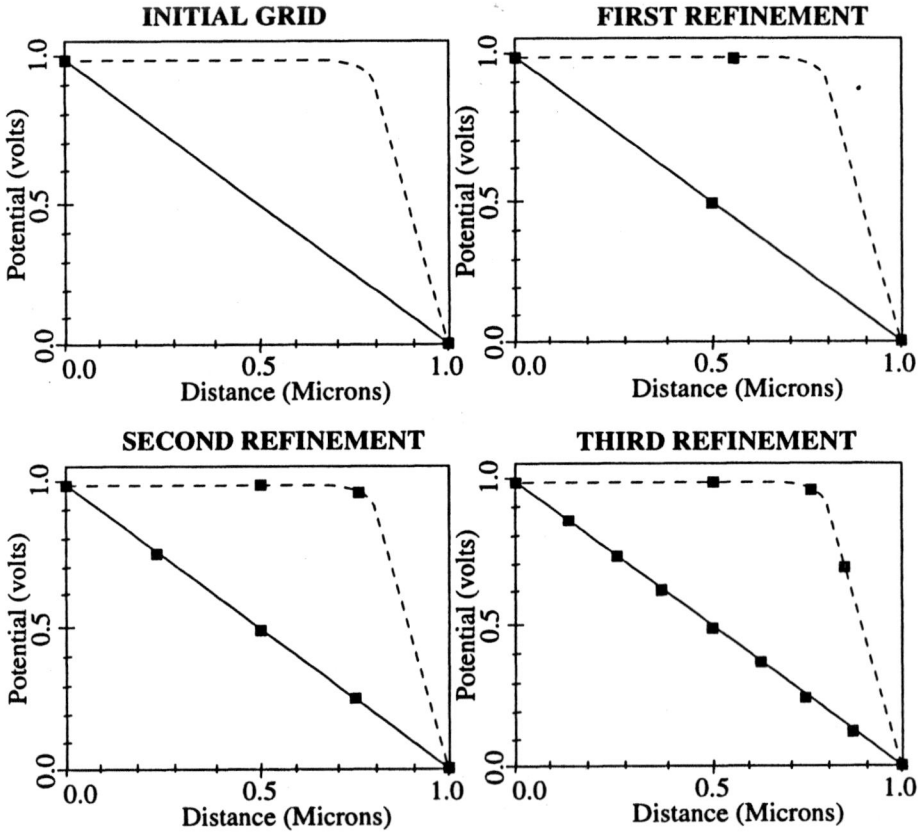

FIGURE 5.1 Interpolated (solid) vs. solution data (dashed) illustrating the importance of performing a solution between each level of refinement.

Adaptive Mesh Refinement

An alternative is to compare solutions in adjacent mesh elements to determine if further refinement is necessary. Herein there are two main options: refine where a particular variable exceeds a value, or refine where the change in that variable across a mesh element exceeds a given value. If several levels of refinement are done in immediate succession, the refinement decisions at higher levels are done using interpolated data [3.2]. The nonlinearity of semiconductor problems makes this inadvisable, i.e., the data used to refine a grid should be updated as soon as possible. Figure 5.1 illustrates this in one dimension. In this hypothetical case, a sharp bend in the potential contour is being refined so that all elements with steps more than 0.1V across them are being refined. With interpolated data, the whole interval would be refined as shown by the solid line in Figure 5.1. If a new solution is performed between levels however, the refinement can detect that the change in potential is localized as in the dashed curve. To allow for this phenomenon, it is necessary to refine one level at a time, performing a new solution between levels. The variable used in the refinement procedure can be any of the key quantities in the problem- in semiconductor device problems it could be potential, electron or hole concentration, electron or hole quasi-Fermi potential, doping, electric field or minority carrier concentration. The value to choose depends on the size of the structure and the accuracy desired. Typically no element would have a step of more than 10-20 kT/q in potential or quasi-Fermi potential across it. Similarly doping wouldn't change by more than two orders of magnitude across an element.

In the simplified semiconductor model, the variable of choice could be the electric field, E or the electric potential Ψ. A user defined refinement criterion can be used with either of these variables if variable accuracy regridding is required. Alternately, inter-element electric field discontinuity can be used as a local error estimator [3.2]. For each mesh element with N_0 neighbours, the error estimator ε_o can be defined as

$$\varepsilon_0 = \frac{\left(\sum_{i=1}^{N_0} |E_0 - E_i|\right)}{N_0}, \tag{5.1}$$

where E_i is the electric field vector in element i. At each adaptive iteration, ε_0 is calculated for each element and the elements are ordered by the calculated ε_0. The largest fraction of them can then be refined by breaking them up into smaller elements.

Since the substrate outside of the active areas has variations in doping density only in the vertical (z) direction and the Dirichlet boundaries are found only in the horizontal

Mesh Generation

(x-y) plane we can use some analytical reasoning to determine the density of grids required prior to actual discretization.

5.2 A Priori Mesh Refinement

The setup of Figure 5.2 consists of a noisy node and a sensitive node separated by a fixed lateral distance. With a fixed number of grids in the x and y directions between these nodes, the peak-peak noise voltage at the sensitive node is determined as a function of the grid density in the z direction for different substrate thicknesses. For the geometry shown in Figure 5.2, noise-coupling is a strong function of grid density in the z direction as expected, since the gradient of the electric field is high in that direction. The number of grids needed (in the z direction) is also a strong function of the substrate thickness.

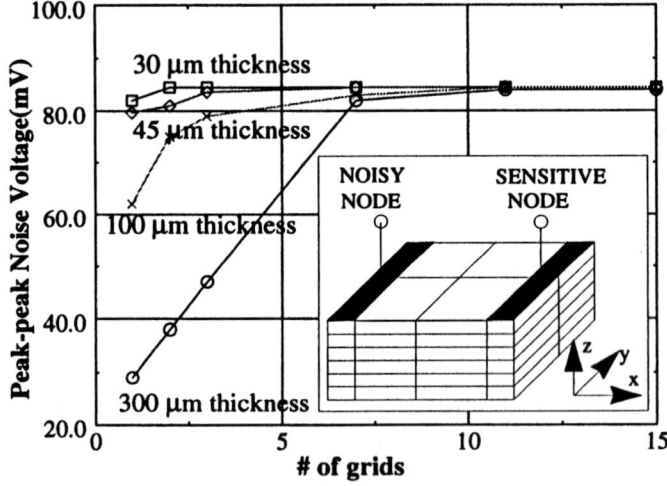

FIGURE 5.2 The peak-peak noise voltage at the sensitive node as a function of the number of grids in the z direction (with the grid density in the x and y directions fixed) for different substrate thicknesses.

A Priori Mesh Refinement

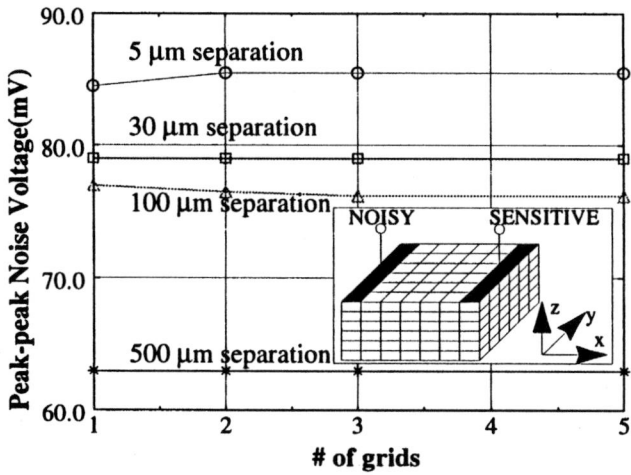

FIGURE 5.3 The peak-peak noise voltage at the sensitive node as a function of the number of grids in the x and y directions (with the grid density in the z direction fixed) for different amounts of lateral separation.

Figure 5.3 shows the effects of grid density in the x and y directions with a fixed grid density in the z direction. Since the fixed boundaries are in the xy plane, the noise coupling is not as sensitive to the number of grids in the x and y directions; consequently, coarse grids are used in these directions. Thus, we empirically determine a grid density based on total substrate thickness from the results of Figure 5.2 and Figure 5.3. Although adaptive refinement is a more common form of mesh generation, in the case of a linear substrate model with contacts/diffusions at the surface or a backside contact, it is relatively easy to determine the density of grids required *a priori* using the technique mentioned above.

Although it is simplest to distribute grids uniformly around a boundary (Dirichlet or Neumann), it is more practical and efficient to place finely discretized grids nearer the boundary and coarser ones as the distance from the boundary increases. Similarly for a substrate with several layers of varying doping densities, it is necessary to use finely discretized grids near the interfaces between the layers as the electric field vector changes there. Another factor to consider in mesh generation is the edge effect. Due to the Neumann boundary at the substrate edges it is necessary to use fine grids there in order to model accurately the change in the magnitude and direction of the

Mesh Generation

Mesh Generation

electric field vector. Once the density of grids required are determined using the empirical techniques mentioned, the distribution of these grids have to be done heuristically. For a monotonically increasing or decreasing mesh spacing size, the following algorithm can be used. Assume that the variable, *number_of_grids*, represents the total number of grids to be distributed in a section of length, *section_length* with minimum and maximum dimensions *section_min* and *section_max* respectively. Firstly, we divide all the grids evenly into *cluster_number* clusters. The locations of the clusters are determined using a geometric progression in the variable *ratio*. Once the clusters are determined, the grids in each cluster, *cluster_grids* are distributed using a geometric heuristic. Note that the dimensions of the clusters increase geometrically only if *ratio* > 2.0.

ratio = (*section_length*)$^{1/cluster\_number}$;

if ratio > 2.0

{

    cluster ( );

    cluster_grids = number_of_grids/cluster_number;

    for 0 < i <= cluster_number

    {

        cluster_min = cluster_location(i-1);

        cluster_length = cluster_location(i) - cluster_location(i-1);

        distribute ( );

    }

}

else

{

    cluster_grids = number_of_grids;

    cluster_length = section_length;

    cluster_min = section_min;

    distribute( );

}

A Priori Mesh Refinement

cluster ( )

{

    cluster_location(0) = section_min;

    cluster_location(cluster_number) = section_max;

    for $0 < i <$ cluster_number

    {

        if monotonically increasing

            cluster_location(i) = section_min + $(ratio)^i$;

        if monotonically decreasing

            cluster_location(i) = section_max - $(ratio)^{cluster\_number - i}$;

    }

}

distribute ( )

{

    if monotonically increasing

    {

        solve for x in

        (cluster_grids - 2) $x^{cluster\_grids}$ - (cluster_grids) $x^{cluster\_grids-1}$ + 2 = 0;

        dx = 0.0;

        count = 0;

        for each grid in cluster

        {

            dx = dx + 2.0 (cluster_length) x^{count} / ((cluster_grids) $x^{cluster\_grids-1}$);

            count + count + 1;

            grid_location = cluster_min + dx;

        }

    }

Mesh Generation

if monotonically decreasing
{
    solve for x in
    $2 x^{cluster\_grids} - (cluster\_grids) x + (cluster\_grids - 2) = 0;$
    $dx = 0.0;$
    $count = 0;$
    for each grid in cluster
    {
        $dx = dx + 2.0 (cluster\_length) x^{count} / (cluster\_grids);$
        $grid\_location = cluster\_min + dx;$
    }
}
}

For a mesh spacing that decreases (increases) from both ends of the section to a minimum (maximum) in the middle of the section, we split the section into two and perform a monotonically decreasing (increasing) distribution in the first half and a monotonically increasing (decreasing) one in the next. By dividing the grid distribution into two phases (*cluster ( ) and distribute ( )*) we are able to better distribute the grids where *section_length* is large.

5.3 Summary

In this chapter adaptive mesh refinement and a priori mesh refinement were discussed as two solutions to the problem of generating the substrate mesh. While the former was shown to be an elegant and effective solution to the problem the latter was chosen for its simplicity and computationally inexpensive applicability. An algorithm for distributing the grids once the density is determined was also presented.

CHAPTER 6
Substrate Modeling in Heavily-Doped Bulk Processes

In Chapter 4 substrate models were developed and discussed that are valid for any process technology in silicon. Additionally, for processes with an epitaxial layer on a heavily-doped bulk (doping density of $10^{18}/cm^3$ or higher) as is typical of many CMOS technologies, a simpler model can be used- the single node model [6.1],[6.2] In this chapter we will discuss the single-node model, its advantages and limitations and some ways to overcome the latter.

6.1 Motivation

Figure 6.1 shows a circuit setup/layout profile that was used with the device simulator, MEDICI to determine the nature of current flow in a heavily doped substrate process (i.e, heavily-doped bulk with lightly doped epitaxial layer) [6.2]. The schematic is similar in nature to that shown in Figure 3.7 and consists of an equivalent drain diffusion to which a switching voltage is applied. The noise sensitive NMOS transistor is placed a distance, *d* away and is biased to ideally produce a constant current. The inductors L1 - L3 represent the parasitic inductances associated with the bond wire and package pins of the die structure and the capacitor C1 represents the capacitance between the backside of the substrate and package cavity (held at ground) through a nonconductive epoxy. The device is taken to be be 200 µm wide giving it an aspect ratio of 100 (W/L = 200/2). The substrate is biased at a potential of -5V and and the

Substrate Modeling in Heavily-Doped Bulk Processes

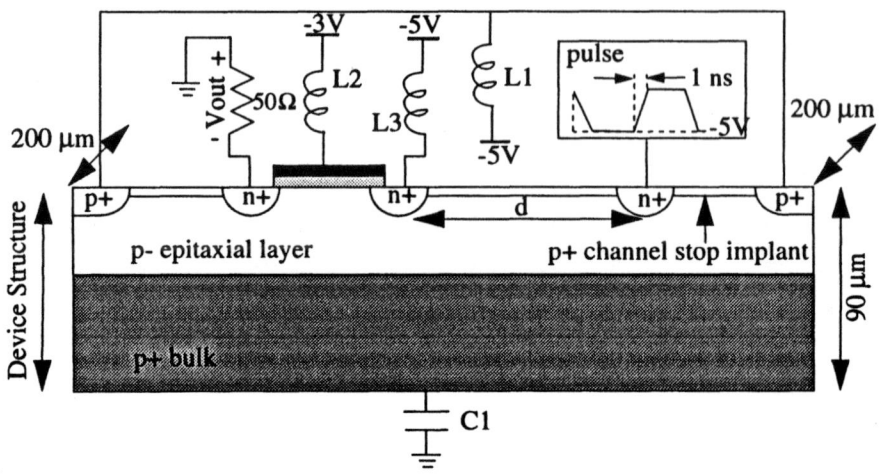

FIGURE 6.1 Circuit setup used to determine current flow in substrate.

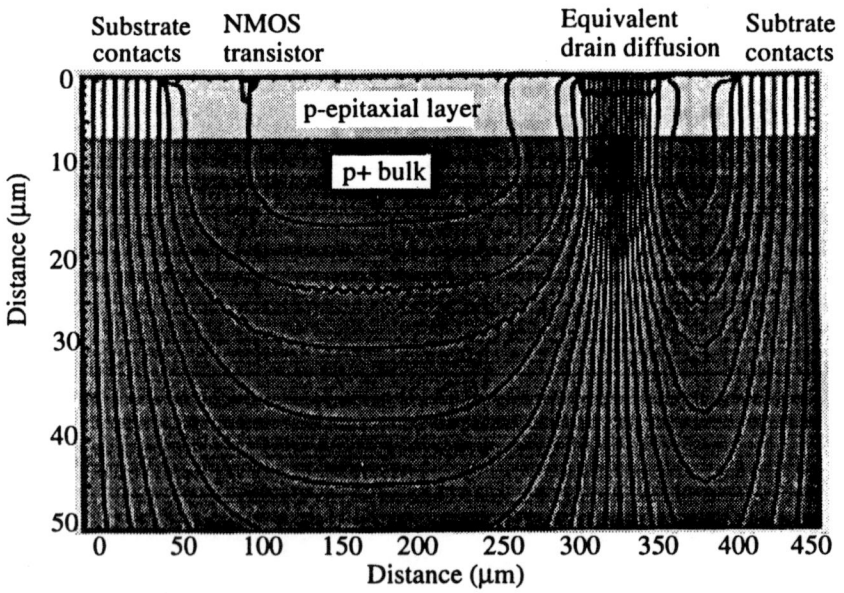

FIGURE 6.2 Current flow lines in a heavily-doped substrate.

Single Node Substrate Model

excitation pulse applied to the equivalent diffusion switches between 0V and -5V with a rise/fall time of 1 ns. Figure 6.2 shows the results [6.2] of a simulation 0.1 ns after the initiation of a 1 ns high-to-low transition at the drain diffusion. The contours represent the current flow lines at intervals of 5% of the total substrate current. From Figure 6.2 it is evident that most of the lateral current from the digital noise source to the sensitive transistor flows through the heavily-doped bulk. Since the bulk is of very low resistivity (0.05 Ω-cm) and the epi (epitaxial layer) is very thin (7 μm) the injected current flows through the epi almost vertically without much lateral spreading and directly into the bulk and then up through the epi again to the substrate contacts. Su etc. [6.2] have corroborated these simulation results with measurements that indicate that the resistance between two contacts in a heavily doped substrate is independent of the distance between them if they are separated by more than four times the effective thickness of the epi layer. This implies that the heavily-doped bulk can effectively be treated as a single node. Consequently any noise injected into it will spread and manifest itself across the entire die.

6.2 Single Node Substrate Model

The equivalent single node model for any given circuit can be developed as shown in Figure 6.3. The transisors and their associated junction capacitances are represented by their appropriate SPICE models while the vertical spreading resistances through the epitaxial layer are modeled by the resistors R_{epi1}-R_{epi3}. The epitaxial resistance can be computed using an analytical formula as a parallel combination of two resistances, R_{AREA} and R_{PER} [6.2]. The area component of the resistance, R_{AREA}, is based on uniform current flow through a rectangular block and is given by

$$R_{AREA} = \frac{\rho T}{A}. \tag{6.1}$$

The parameters ρ and T represent the resistivity and effective thickness of the epitaxial layer, respectively and A is the surface area of the active area. The resistance due to current flow at the perimeter of the acitve area, R_{PER}, is based on uniform conduction through a hemisphere and is given by

$$R_{PER} = \frac{\rho}{P} \tag{6.2}$$

where P is the perimeter of the active area. The effective spreading resistance is the parallel combination of (6.1) and (6.2) and is given by

$$R_{EPI} = \left(\frac{k_1 \rho T}{(L+\delta)(W+\delta)}\right) \Big\| \left(\frac{k_2 \rho}{2(W+L+2\delta)}\right) \quad (6.3)$$

where the variables k_1, k_2, and δ are parameters that fit the formula to empirically measured resistances. Su etc. have reported that the results from (6.3) with $k_1 = 0.96$, $k2 = 0.71$, and $\delta = 5.0 \mu m$ are within 15% of measured results.

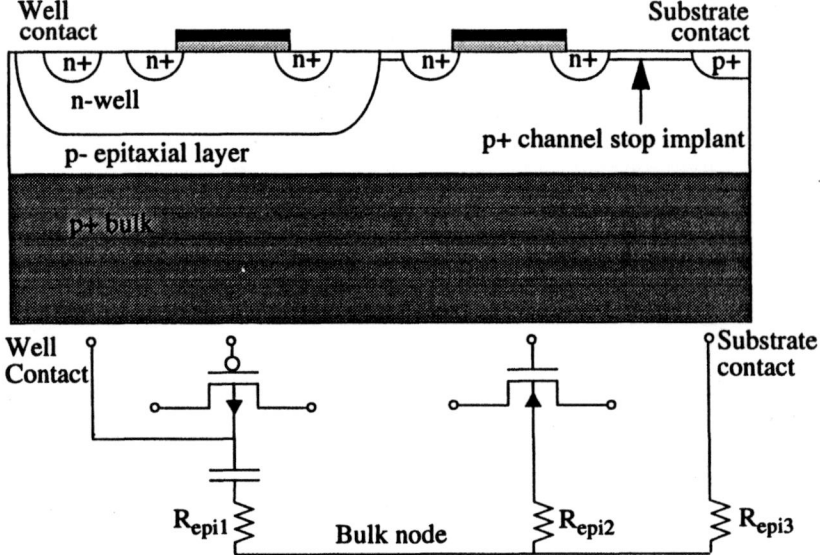

FIGURE 6.3 The single node model for a heavily-doped substrate.

Figure 6.4 compares the simulation results using the single node model with measured results from the experimental chip described in Section 4.10.2 and also with simulation results using the macromodel. It plots the peak-peak voltage and settling time of the noise waveform at a sensitive transistor as a function of the number of package pins (10 nH per package pin bond wire) used to bias the substrate. It is apparent from Figure 6.4 that the single node model quite accurately represents coupling in the substrate.

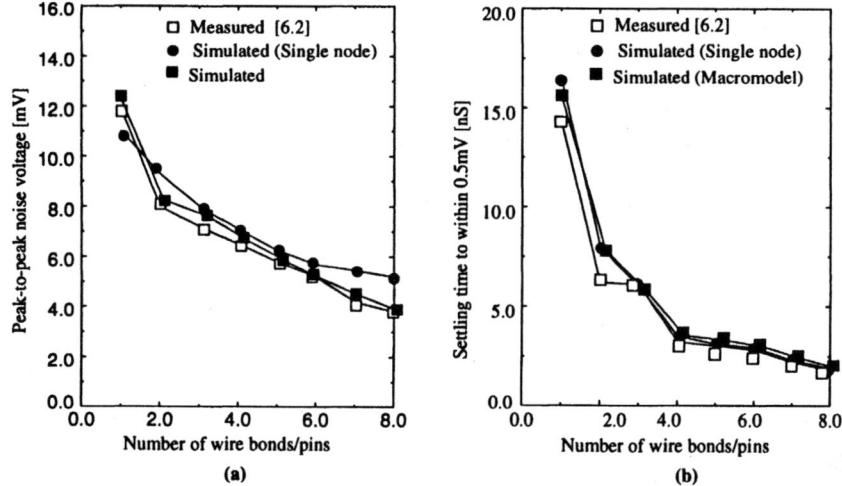

FIGURE 6.4 Comparison of (a) peak-peak and (b) settling time behavior of the noise voltage as a function of the number of bonding pads used to bias the substrate in the experimental chip of Section 4.10.2.

6.3 Modified Single Node Substrate Model

Note that in the aforementioned single node substrate model lateral current flow between transistors/contacts/wells in the epitaxial layer has been ignored and this is justifiable as long as the distance between them is larger than four times the thickness of the epitaxial layer. Unfortunately, this is a major limitation of this model. Lateral resistance becomes significant when it is comparable in magnitude to vertical resistance to the bulk and this is particularly (and typically only) the case when an active area is surrounded by a guard ring. By ignoring lateral current flow we also assume that all the current in the substrate flows directly from the active area in consideration to the bulk, which is incorrect in these cases. Consequently, the epi resistance computed by (6.3) will underestimate the true resistance to bulk. To illustrate the problem consider the set up of Figure 6.5. The noise injected by a number of inverters capacitively coupled to the substrate is measured at the drain of a sensitive transistor which is protected by a p+ guard ring surrounding it a distance of 6 μm away.

Substrate Modeling in Heavily-Doped Bulk Processes

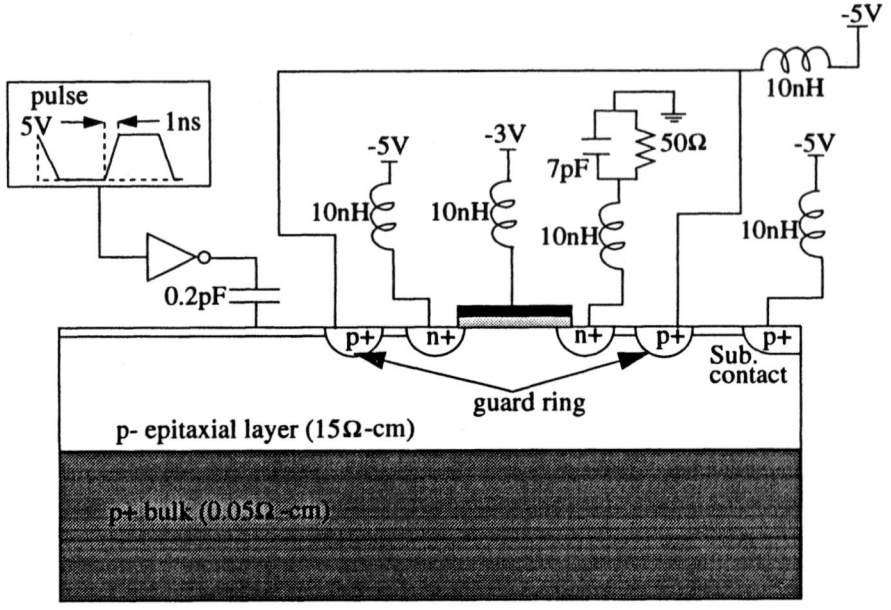

FIGURE 6.5 Experimental setup to demonstrate limitation of single node substrate model.

The single node model for the substrate of Figure 6.5 can be represented as shown in Figure 6.6.

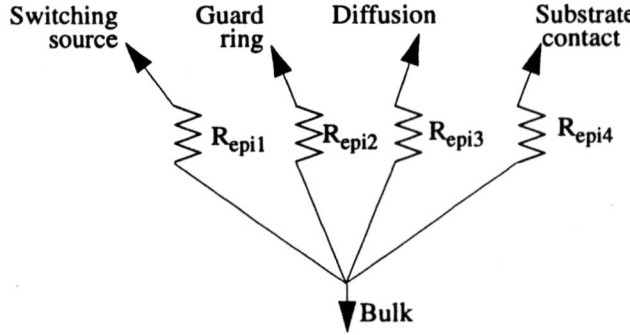

FIGURE 6.6 Single node model for substrate of Figure 6.5.

Modified Single Node Substrate Model

In reality, the model of Figure 6.6 is incorrect since it ignores the lateral current flow between the guard ring and the sensitive transistor. This however can easily be overcome by introducing an additional resistance representing the lateral coupling between them as shown in Figure 6.7.

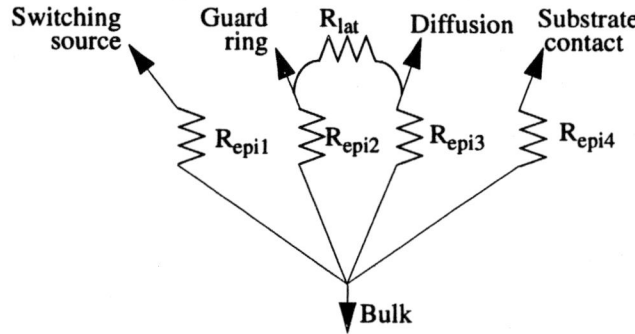

FIGURE 6.7 Modified single node model incorporating lateral current flow.

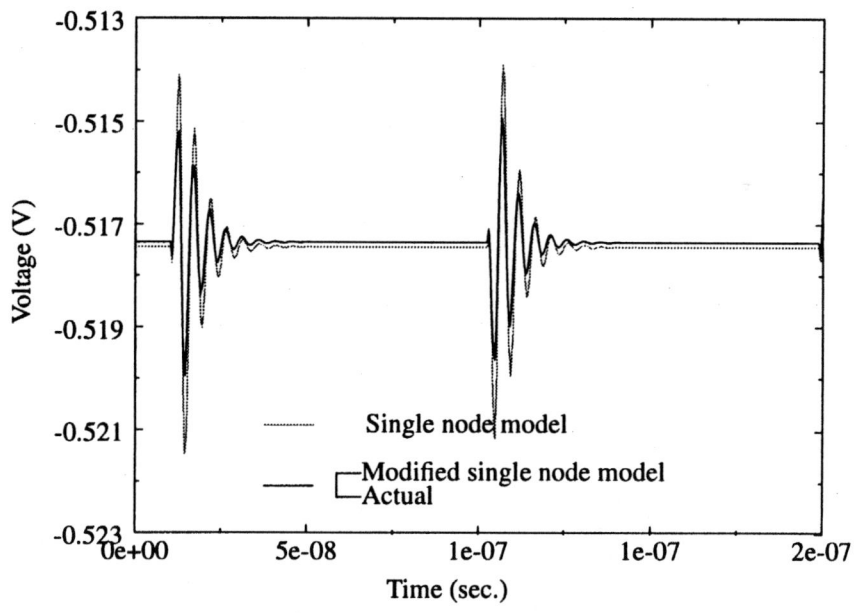

FIGURE 6.8 Comparison of simulation results between the single node model and the modified single node model.

Substrate Modeling in Heavily-Doped Bulk Processes

Figure 6.8 shows the noise voltage at the drain of the sensitive transistor as a result of coupling through the substrate from the switching sources. A comparison of the simulation results using the single node model and the modified single node model shows that by ignoring lateral coupling between the guard ring and the diffusion, the single node model underestimates by almost 40% (in peak-to-peak magnitude) the effectiveness of the p+ ring in shielding the transistor from the noise.

Now that we know that lateral current flow is an important part of the substrate coupling model for guard ring shielded active areas the next question that arises is how do we determine the corresponding resistances? Going back to the dc macromodeling technique described in Chapter 4 we see that if we include the backside of the die as a port, the admittance macromodel of the given layout/circuit will consist both of a column of admittance parameters that represents the coupling between the active areas and the bulk and also parameters that represent the lateral coupling between the active areas. For the example of Figure 6.7 the admittance macromodel would be as follows:

$$\begin{array}{r} \text{Switching source} \\ \text{Guard ring} \\ \text{Diffusion} \\ \text{Substrate contact} \\ \text{Bulk} \end{array} \begin{bmatrix} y_{s-s} & y_{s-g} & y_{s-d} & y_{s-c} & y_{s-b} \\ y_{g-s} & y_{g-g} & y_{g-d} & y_{g-c} & y_{g-b} \\ y_{d-s} & y_{d-g} & y_{d-d} & y_{d-c} & y_{d-b} \\ y_{c-s} & y_{c-g} & y_{c-d} & y_{c-c} & y_{c-b} \\ y_{b-s} & y_{b-g} & y_{b-d} & y_{b-c} & y_{b-b} \end{bmatrix}$$

with columns labeled: Switching source, Guard ring, Diffusion, Substrate contact, Bulk.

where y_{s-b} ($= y_{b-s}$) represents the conductance between the switching source and the bulk, y_{d-b} ($= y_{b-d}$) the conductance between the diffusion and bulk and so on. The other non-diagonal entries in the admittance matrix represent lateral coupling between the various ports. For the example of Figure 6.7, R_{lat} will therefore be given by $1/y_{g-d}$ ($= 1/y_{d-g}$).

Since the numerical solution procedure entailed in the dc macromodeling technique is computationally cumbersome especially as the size of the circuit to be simulated increases, it should be employed only in cases where an active area is closely surrounded by a guard ring. In such a case, an artificial Neumann boundary (ie, current

Summary

flowing across the boundary is zero) can be drawn around the guard ring since we know that lateral current flow is significant only between the guard ring and the active area it surrounds. Thus, only the local area that includes the transistor and the guard ring needs to be numerically simulated to determine the corresponding vertical resistances to bulk and lateral resistance. In all other regions of the substrate (i.e, where lateral current flow can be ignored) an analytical formula similar to (6.3) can be employed to determine the vertical spreading resistance to the bulk. For different processes, the fitting parameters k_1, k_2 and δ in (6.3) will have to be recomputed and this can be done as a preprocessing step by suitable curve fitting onto a set of data points. A sample of these data points can be obtained by repeated numerical simulations with active areas of different sizes. Each numerical simulation is computationally quite inexpensive and involves a single ICCG iteration on a matrix that is small since we are dealing with only one active area at a time.

6.4 Summary

In this chapter, the single node substrate model was introduced as a strategy to simulate substrate coupling in processes with heavily-doped bulks and lightly doped epitaxial layers. Some limitations of this model and means to overcome them were discussed.

The single node model is more efficient than the dc macromodel discussed in Chapter 4 in that it requires a lot fewer resistances, but care has to be taken to incorporate lateral resistances when required. In regions where lateral current flow is important numerical simulation can be used to determine corresponding resistance values. Elsewhere an analytical formula can be employed to determine vertical spreading resistance to the bulk. The analytical formula can be developed as a preprocessing step using data points collected from several numerical simulations.

REFERENCES

[6.1] T.J. Schmerbeck, R.A. Richetta, and L.D. Smith, "A 27 MHZ mixed A/D magnetic recording channel DSP using partial response signalling with maximum likelihood detection," *Technical Digest of the International Solid State Circuits Conference*, pp. 136-137, Feb. 1991.

[6.2] D.K. Su, M.J. Loinaz, S. Masui and B.A. Wooley, "Experimental Results and Modeling Techniques for Substrate Noise in Mixed-Signal Integrated Circuits," *IEEE Journal of Solid State Circuits*, vol. 28, no. 4, April 1993.

CHAPTER 7

Substrate Resistance Extraction for Large Circuits

In Chapter 6 we discussed a modeling strategy for substrates with a heavily-doped bulk and a lightly doped epitaxial layer. The advantage of the single node model for such substrates is that, once characterized for a particular process it can readily be used for any circuit structure that uses the same process, irrespective of size. Unfortunately the model does not extend to processes which do not have a heavily-doped bulk which is in effect a collector and distributor of stray electrons and holes. On the other hand, the dc macromodeling technique of Chapter 4 can be used for any Silicon process. The problem with the latter approach however, is that it becomes computationally expensive as the size of the circuit to be simulated increases. In this chapter we will introduce techniques that allow for the numerical computation (extraction) of substrate resistance for large circuit structures.

7.1 Nested Macromodeling

This technique was adopted by Johnson, Knepper, et. al [7.1] as a sloution to studying a disturb problem encountered with a high-speed bipolar array chip. As described in Section 4.2, each control volume in the semiconductor substrate can be modeled as shown in Figure 7.1. Inherent to this approximation is the assumption that potential is constant over each of the volume element's faces and that current is uniform in any coordinate direction between parallel faces. The substrate mesh is obtained by placing many of these control cells side by side and the key to obtaining an accurate model rests in how one builds it. In modeling the substrate, care must be taken to ensure that the cell size used is small enough to permit accurate modeling in the vicinity of devices of interest. The grid of interconnection points must be sufficiently detailed to permit devices to electrically couple to the substrate without forcing too much current to be injected at any one point.

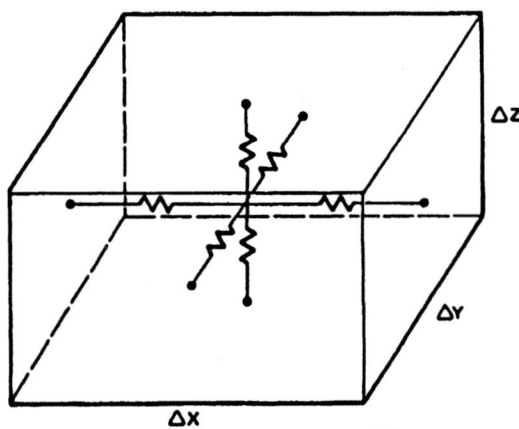

FIGURE 7.1 **Unit cell model for a small block of substrate material.**

In order to properly model local coupling between adjacent transistors, the top layer(s) in the substrate distributed model must include a reasonably fine resolution of resistances. Since the currents that flow deep within the substrate generally flow distances much greater than the chip thickness, the bottom layer(s) can be adequately modelled by a much coarser network. By reducing the detail of modeling as we move deeper into the substrate, we also reduce the size of the resistive network that we ultimately must analyze. A possible approach to the modeling procedure is shown in Figures 7.2-7.4. Figure 7.2 shows 64 cells placed side-by-side to form a larger structure, which we will call the "A" level unit cell. Taking four of these "A" level cells and combining it with another unit cell having twice the dimensions of the "A" level cell, the structure of Figure 7.3 can be constructed. The above procedure can be repeated for as many levels of modeling as desired. For example, four "B" level cells combined with one "C" level cell gives the structure of Figure 7.4 and so on.

Nested Macromodeling

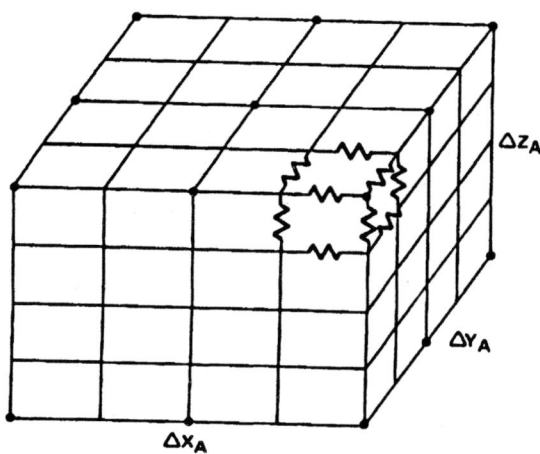

FIGURE 7.2 Representation of the "A" cell used to build the complete substrate model.

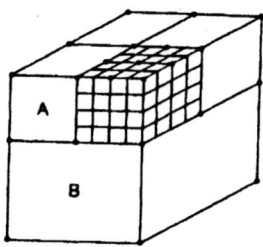

FIGURE 7.3 Four "A" cells and one "B" cell nested together.

As the various control cells are combined, the total number of resistances and nodes increases. However, in general, the only nodes that are necessary in going from one model to the next as we nest the control cells are nodes along the surface of the resulting structure. It is these nodes and not ones in the interior, that must be retained in order to form the next larger structure or model. If we reduce each model as it is created so that all the interior nodes and elements are combined, then at each stage in the model building process only the minimal number of elements required to accurately describe the substrate will be retained.

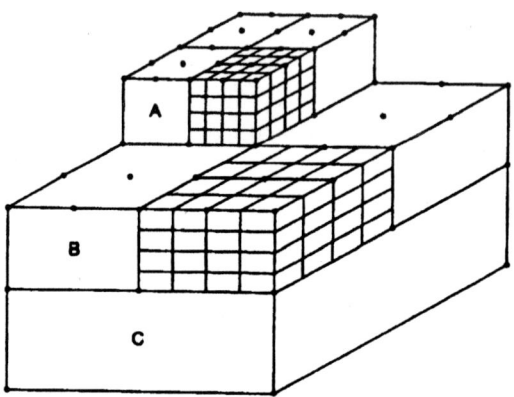

FIGURE 7.4 Interconnection of "A", "B", and "C" cells.

Each of the control cells previously described may be represented by Y_p, an $n \times n$ admittance matrix and when any two control cells are connected together, the resulting structure can be represented by a new matrix that contains all of the elements of the two separate matrices. In order to reduce the matrix, the dc macromodeling technique of Section 4.8 can be utilized. Let the matrix Y_p be created such that the first k rows correspond to the k external nodes of the reduced network. Included with external nodes are all nodes on the surface of the control cell that are needed to connect this model to form the next larger structure. Also included with the external nodes are all points on the top surface that are needed to couple the substrate to the circuitry above it. The matrix Y_p can then be partitioned into four regions:

$$\begin{bmatrix} Y_{kk} & Y_{ik} \\ Y_{ik} & Y_{ii} \end{bmatrix} \begin{bmatrix} V_k \\ V_i \end{bmatrix} = \begin{bmatrix} J_k \\ J_i \end{bmatrix} \tag{7.1}$$

where the index i represents the internal nodes. Y_{kk} is a $k \times k$ submatrix, which is an exact equivalent conductance representation of the k external nodes in the original network. The original matrix Y_p may be replaced by the smaller $k \times k$ submatrix without any loss of accuracy. The modeling process therefore consists of building a composite substrate model by connecting many control cells together. As the control cells are nested, the resulting model is reduced using the dc macromodeling technique so that only those nodes that are aboslutely necessary for constructing the next level of the model are retained. When completed, the final model will be an exact equivalent for the large resistive network, which could be composed of hundreds of thousands of discrete resistive elements.

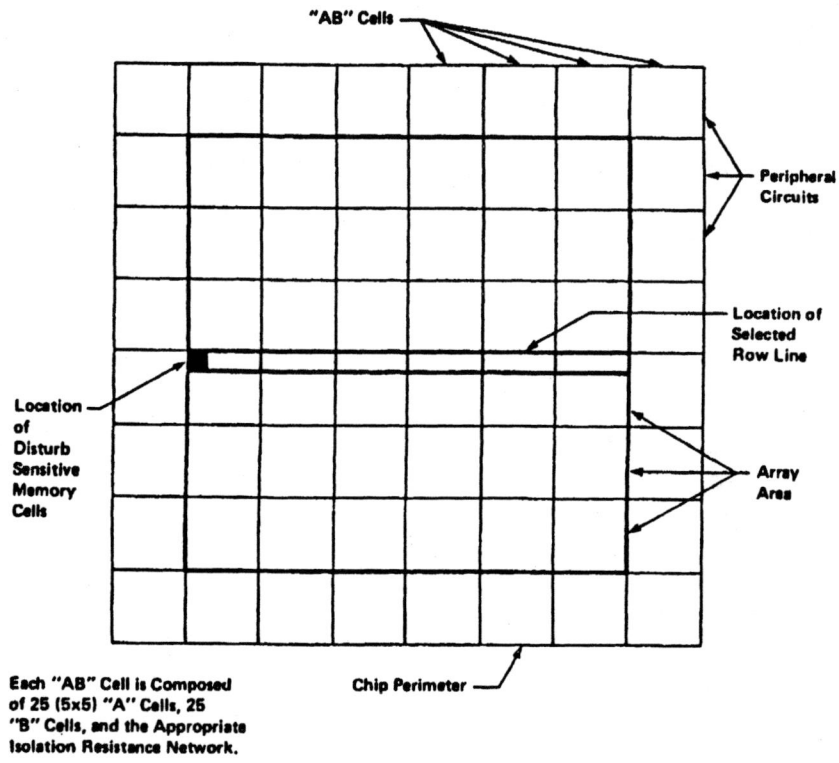

FIGURE 7.5 Top view of the chip substrate showing "AB" cell locations.

The nested macromodeling approach was employed in the simulation of disturb failures in a bipolar array memory chip [7.1]. The array cells on a switching row line in

Substrate Resistance Extraction for Large Circuits

the chip exhibited a reduced cell noise margin (internal node-to-node voltage) during selection. If under certain conditions of supply voltage, temperature, address pattern, and device parameter variations, the cell node voltage differences collapsed to zero, the cell could switch states thus exhibiting a disturb failure. Figure 7.5 shows the top view of the chip, and the locations of the "AB" cells used for modeling. Without reduction, the total chip model would have contained over 400,000 nodes which would have made fast matrix solution next to impossible. In order to verify the nesting approach, the standard silicon wafer four-point probe measurement of sheet resistance was simulated. A number of vertical tie points were made in the chip to provide current paths to the substrate bulk in the four point probe measurement. Consequently, a number of north-south and east-west points were selected to conform to measurement probe locations in the array chips. A comparison of the point-to-point simulated resistance with measured resistance is shown in Table 7.1.

TABLE 7.1 Comparison of point-point simulated substrate resistance values with measurements. (All values in ohms, measured values + or - 10 percent)

Chip Locations		Calculated	Measured
East-West:			
	A1-B1	416	425
	A2-B2	402	395
	A3-B3	404	375
	A4-B4	420	390
	A5-B5	433	403
North-South:			
	C1-D1	783	602
	C2-D2	709	516
	C3-D3	627	504
	C4-D4	591	482
	C5-D5	459	351

The results indicate that the simulated resistance values are extremely accurate for the east-west measurements but less so for the north-south cases. This is a result of the fact that the chip has isolation diffusion lines running east-west which not only carries a large portion of the substrate current but is also accurately modelled being atop the substrate model. For the north-south cases most of the current must flow is into the substrate and since the modeling there is less precise the resulting simulated resis-

Interpolated Macromodeling

tance values accumulate error. It is apparent from the results that while this technique allows for the computation of some large circuits its accuracy depends on the level of detail of the model used. Increasing the accuracy of the model however will lead to an increase in the cpu time required.

7.2 Interpolated Macromodeling

A direct extension of the dc macromodeling technique to large circuits requires impractical amounts of computer memory and CPU time. One solution to this problem is to divide the substrate mesh into partitions which can be macromodeled independently of each other and then put together in a nested fashion. A possible partitioning scheme is shown in Figure 7.6. where the partitions are generated by slicing through the substrate in the vertical direction.

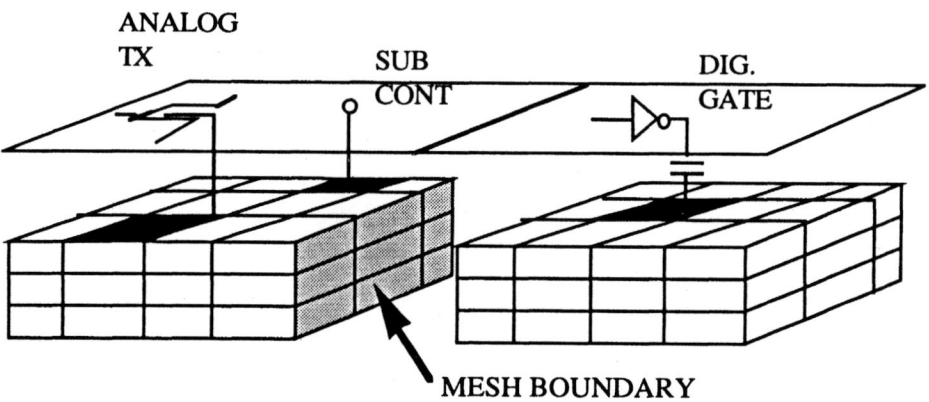

FIGURE 7.6 A partitioning scheme for the substrate mesh.

The two partitions of Figure 7.6 are combined together by connecting every node on the mesh boundary of one with the corresponding one on the other. The partitioning scheme is similar to that of the nested macromodeling scheme of Section 7.1. The advantage of the latter scheme is that there is no loss of accuracy in the modeling process since every node on the mesh boundaries of two partitions are connected together. The disadvantage of it however is the computational overhead involved. Since every node on the mesh boundary is now an external node (as defined in Sec-

Substrate Resistance Extraction for Large Circuits

tion 7.1) the macromodel of each partition is larger than in the previous case and the computation of it is more cumbersome. Also, since the number of nodes on the mesh boundaries of adjacent partitions must be identical, the meshing in one partition affects all neighbouring ones. Hence mesh lines necessary to provide a fine discretization in one partition will spill over into adjacent partitions where a fine discretization may not be necessary.

The potential on the mesh boundary varies approximately linearly with distance which suggests that that only a fraction of the nodes need be defined as external nodes/ports [7.2]. An interpolating polynomial can be used to compute potentials on the remaining nodes. In computing the Y-parameters of the macromodel, only one port is connected to a voltage source of unit value (with all others grounded) at a given time. Defining only the four corner nodes of the mesh boundary as external nodes/ports (Figure 7.7), a linear Lagrangian potential can be used to approximate the potential on the remaining nodes [7.3].

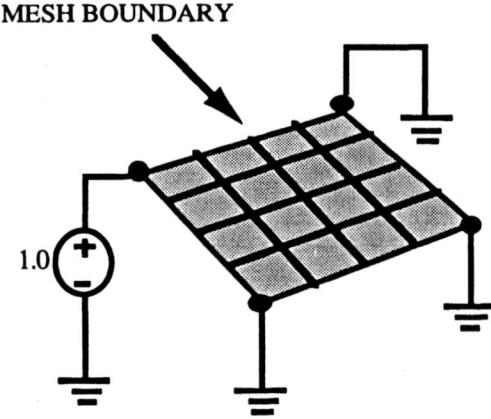

FIGURE 7.7 Mesh boundary with four corners defined as external nodes/ports.

Hence the potential at any given node is given by

$$V = K\left(1 - \frac{x}{X}\right)\left(1 - \frac{y}{Y}\right) \quad (7.2)$$

Interpolated Macromodeling

where x and y are the horizontal and vertical distances from a corner node voltage source of value K, and X and Y are the horizontal and vertical dimensions of the partition. Figure 7.8 shows values of the interpolating polynomial with K=1.00, for a mesh boundary with 25 nodes.

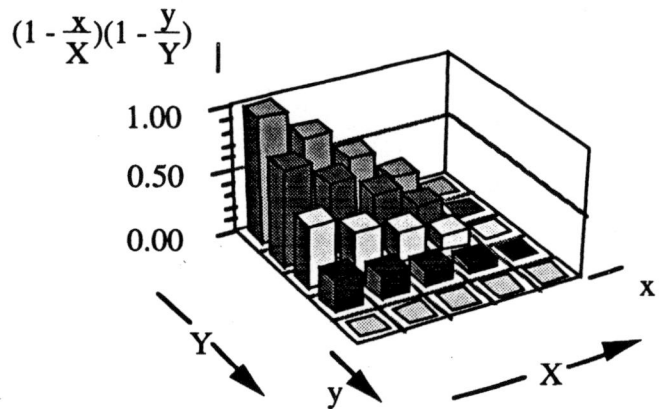

FIGURE 7.8 Lagrangian interpolation on the mesh boundary.

After the potentials on the boundary nodes are fixed, the current flowing across the boundary (i.e, the current flowing through every boundary node) is computed. Since we have defined only the four corner nodes as ports, the boundary currents have to be divided between these four ports. At each corner a Lagrangian interpolating polynomial is used to define scale factors that determine the contribution of each boundary node current to that particular corner as shown in Figure 7.9. The scale factor for a boundary node is a function of its position with respect to the corner in consideration. Moreover, at every boundary node the sum of the scale factors associated with the four corners add up to one. Thus in each partition, rather than determining the Y-parameters for each node on the mesh boundary, the macromodeling problem is reduced to solving for the Y- parameters of the four corners of the boundary. (Note that all nodes that connect the substrate to the electrical circuit must additionally be included in the macromodel.)

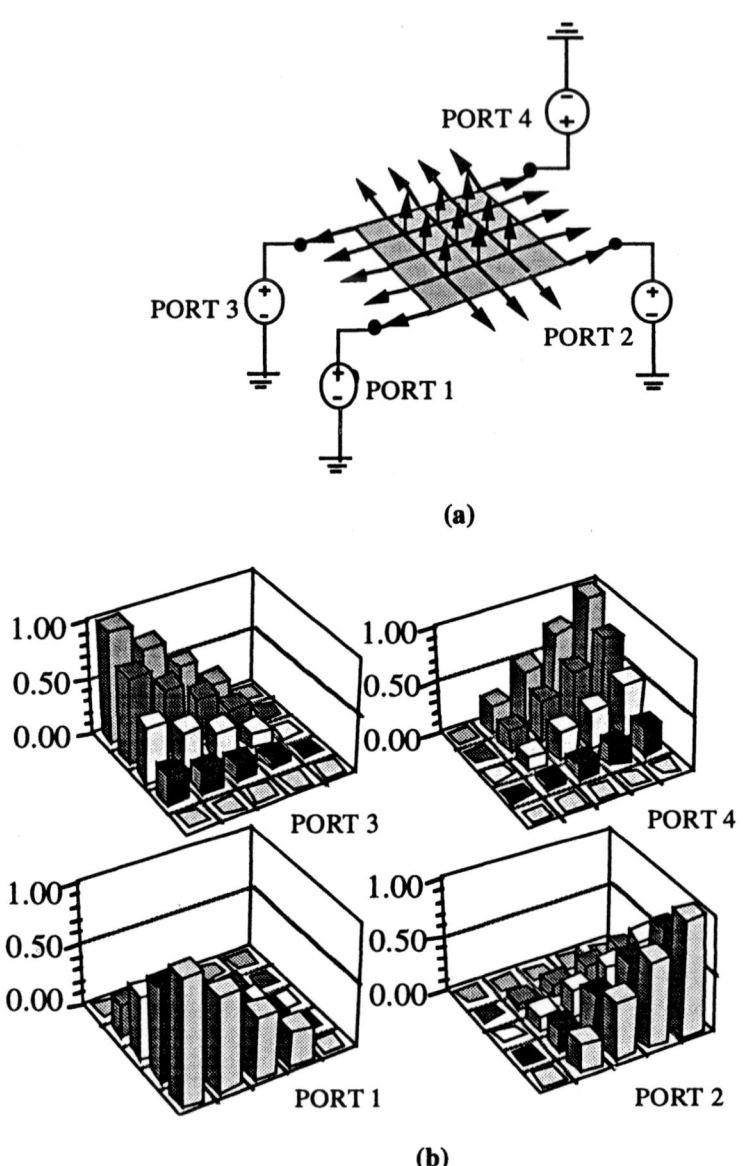

FIGURE 7.9 (a) Boundary currents and (b) Scale factors associated with each port for the boundary currents.

Interpolated Macromodeling

As the macromodels for the partitions are solved for, they are also combined in a nested fashion so that at every stage in the process all internal nodes are eliminated. The end result of this modeling procedure is a macromodel consisting only of those nodes which connect the substrate to the electrical circuit and this macromodel is equivalent to the large substrate mesh consisting of hundreds of thousands of resistors that we started off with. Although a linear interpolating polynomial has been used to describe the partitioning technique, a higher order polynomial can be used to obtain higher accuracy at the expense of a longer computational time.

Figure 7.10 shows a circuit used to verify the interpolated partitioning technique. It consists of an inverter with its output capacitively coupled to the substrate, a sensitive current source, and two subsrate contacts. The simulation results for the drain voltage of the NMOS transistor without partitioning and with four partitions are displayed in Figure 7.11. Some reported measured results from a test chip [7.4] have also been used to verify this technique. The test chip consists of a block of CMOS inverters capacitively coupled to the substrate. The noise injected into the substrate is measured by resistively loaded NMOS transistors distributed across the chip, some of which are shielded from the noise using guard rings placed at varying distances from them and biased either with a dedicated package pin or connected to a large substrate contact. For simulation purposes, the chip was divided into 80 partitions. The measured and simulated results for different cases are compared in Figure 7.12.

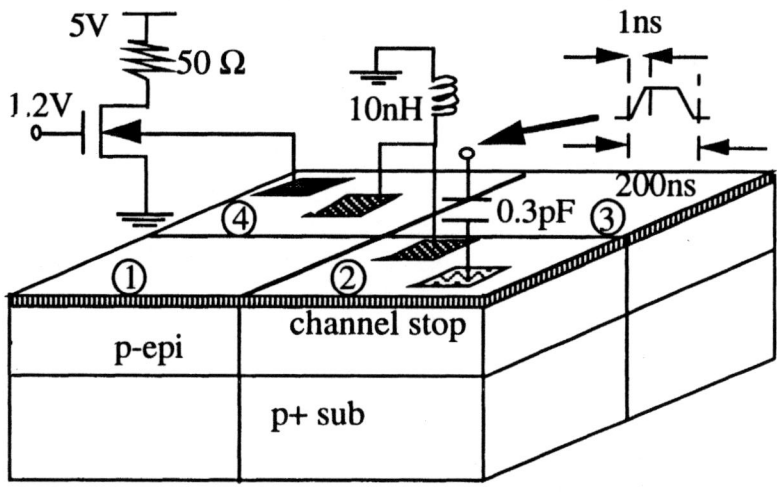

FIGURE 7.10 Example circuit divided into four partitions.

Substrate Resistance Extraction for Large Circuits

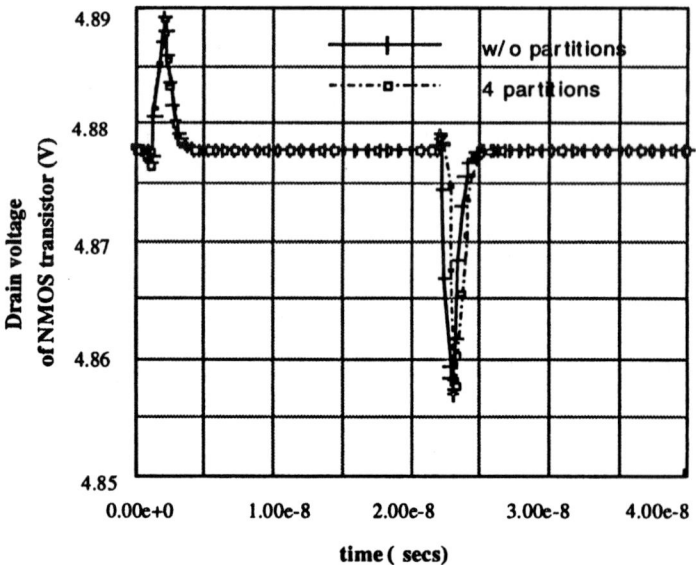

FIGURE 7.11 Drain voltage of NMOS transistor in Figure 7.10.

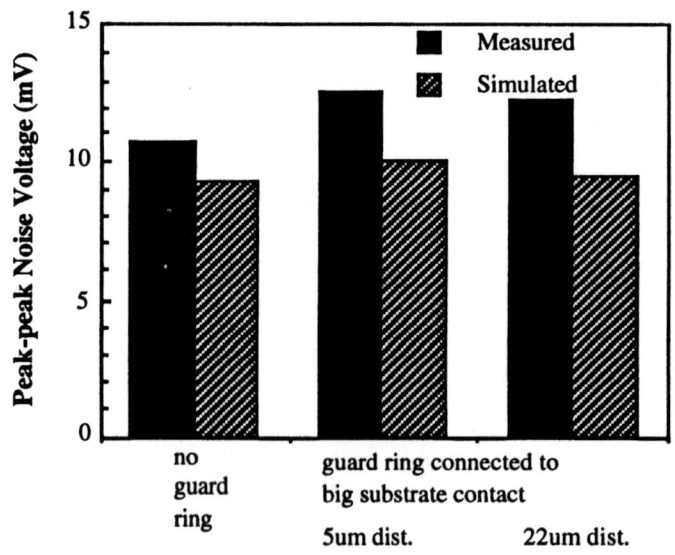

FIGURE 7.12 Partitioned simulation results vs. measured results.

Summary

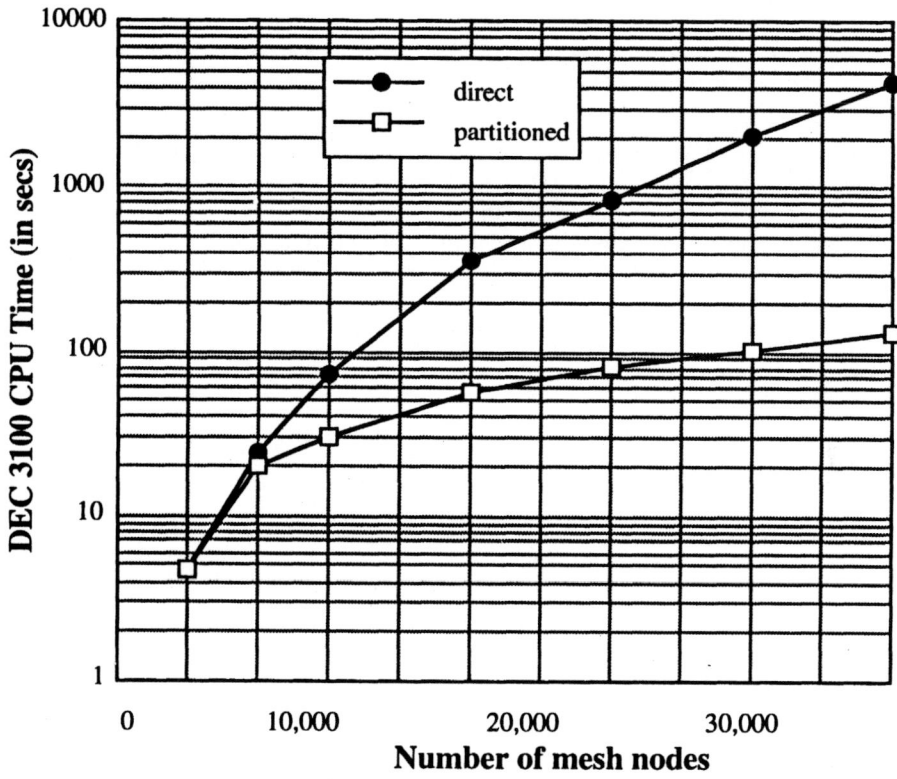

FIGURE 7.13 Direct vs. Partitioned macromodeling.

Figure 7.13 compares the CPU time required for the dc macromodeling technique using the direct approach (of Section 4.8) and the (interpolated) partitioned approach. For more than 40,000 nodes in the subtrate mesh, there is inadequate computer memory to perform the simulation without partitioning.

7.3 Summary

In this chapter we introduced two techniques to handle the large number of mesh nodes in the substrate model for the extraction of substrate resistances.

Both techniques involve partitioning the substrate, developing an independent macromodel for each partition and then connecting them together in a nested fashion. In both cases partitions are connected together through a selected number of nodes called external nodes. When connecting two partitions together, in the nested macromodeling approach, the remaining nodes on the mesh boundary (other than the external nodes) are ignored while in the interpolated macromodeling scheme the remaining nodes are assumed through interpolation. Both techniques provide approximate solutions to the problem but the accuracy of the solution procedure can be controlled by changing the number of external nodes used while partitioning.

In the remainder of the book we will address circuit design and layout techniques to minimize or overcome the substrate coupling problem and also other sources of mixed-signal coupling in integrated circuits.

REFERENCES

[7.1] T.A. Johnson, R.W. Knepper, V. Marcello, and W. Wang, "Chip Substrate Resistance Modeling Technique for Integrated Circuit Design," *IEEE Transactions on Computer-Aided Design*, vol. CAD-3, No. 2, April 1984, pp. 126 - 134.

[7.2] N.K. Verghese and D.J. Allstot, "Simulation of Substrate Coupling in Mixed-Mode VLSI Circuits," SRC-CMU CAD Center Tech. Report No. CMUCAD 93-02.

[7.3] E.J. Bracken, private communication.

[7.4] D.K. Su, M.J. Loinaz, S. Masui and B.A. Wooley, "Experimental Results and Modeling Techniques for Substrate Noise in Mixed-Signal Integrated Circuits, " *IEEE Journal of Solid State Circuits*, vol. 28, no. 4, April 1993.

CHAPTER 8
Modeling Chip/Package Power Distribution

8.1 Effect of Power Bus Structure on Noise coupling

The power bus that interconnects the switching and non-switching functions on the chip is a major source of coupling between widely separated circuits. Switching return currents take the path of least impedance which is often an on-chip path through the power rails. The presence of on-chip decoupling capacitance distributed on the bus can lessen the amount of bus fluctuation due to these switching currents. The simple discussion that follows assumes that the bus is routed on a metal level but busses are often routed or strapped by salicided polysilicon, silicon, or diffused areas as well. Figure 8.1 shows one rail of a power bus in a random or tree power grid structure. There are two switching functions and two non-switching or quiet circuit functions. There is a single chip pad for the supply rail for both switching and non-switching functions. There is on-chip resistive bus drop which provides coupling between switching and non switching functions in addition to the inductive power bounce due to the package inductance connected to the chip power pad. The additional on-chip resistive coupling, which can be substantial, makes this approach usually undesirable. Sometimes additional bus resistance can be beneficial in specific area where it tends to damp RLC ringing of the power bus. The tree power bus scheme usually requires minimum metal area for the bus but must be hand crafted for each design. There is at least one automatic power supply bus routing tool available that routes the bus based on analog signal integrity constraints.[8.13]

Modeling Chip/Package Power Distribution

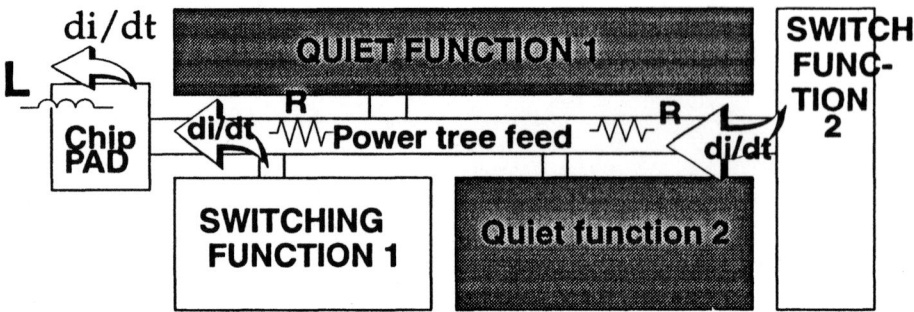

FIGURE 8.1 RANDOM OR TREE (not recommended)

Figure 8.2 shows a STAR connection of a power rail between the switching and non-switching functions to the common chip pad. This scheme provides the minimum coupling prior to hook-up to the common pad but suffers from high bus resistance. This scheme requires more metal hook-up area than the TREE.

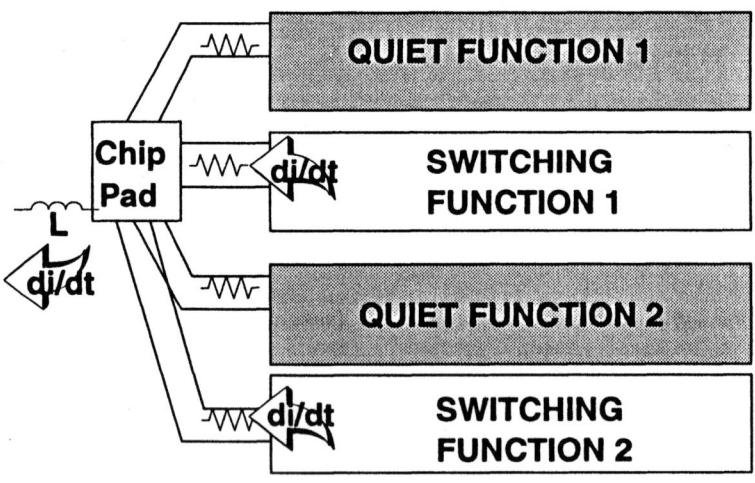

FIGURE 8.2 STAR (maximum isolation to pad)

Effect of Power Bus Structure on Noise coupling

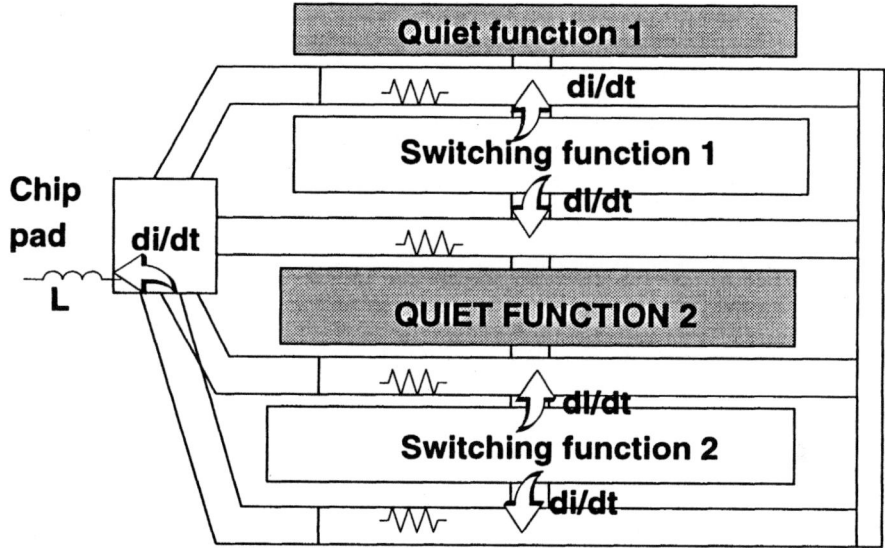

FIGURE 8.3 Simple Grid (General Bus design)

Figure 8.3 shows a GRID power connection scheme of a power rail. This scheme can minimize the on-chip resistance term by providing multiple paths for the current flow by providing a cross hatched grid of metal on multiple connection levels. It also provides a convenient structure to accommodate on-chip decoupling capacitance. Because of the low bus resistance the decoupling capacitance can be located in almost any area within the GRID bus. This scheme does not provide isolation as good as a STAR hook-up without this added chip decoupling capacitance. It does have the advantage that it is a general purpose bus that does not need to be custom for each design. This bus does take up more chip metal area than the STAR or TREE bus hook-up schemes. Figure 8.4 shows a simple two level metal power grid structure from the paper of [8.4]. This particular power grid is implemented on a chip that has dedicated wiring channels between circuit cells. Some of this dedicated wiring area is used to strap the power supply rails together in a cross-hatched grid to lower the bus impedance. This is done in the analog area where bus impedance is most important. The CMOS logic power bus is not gridded since the pure cmos logic has much more

Modeling Chip/Package Power Distribution

noise immunity and generally requires more metal area be reserved for interconnect wiring versus power bussing.

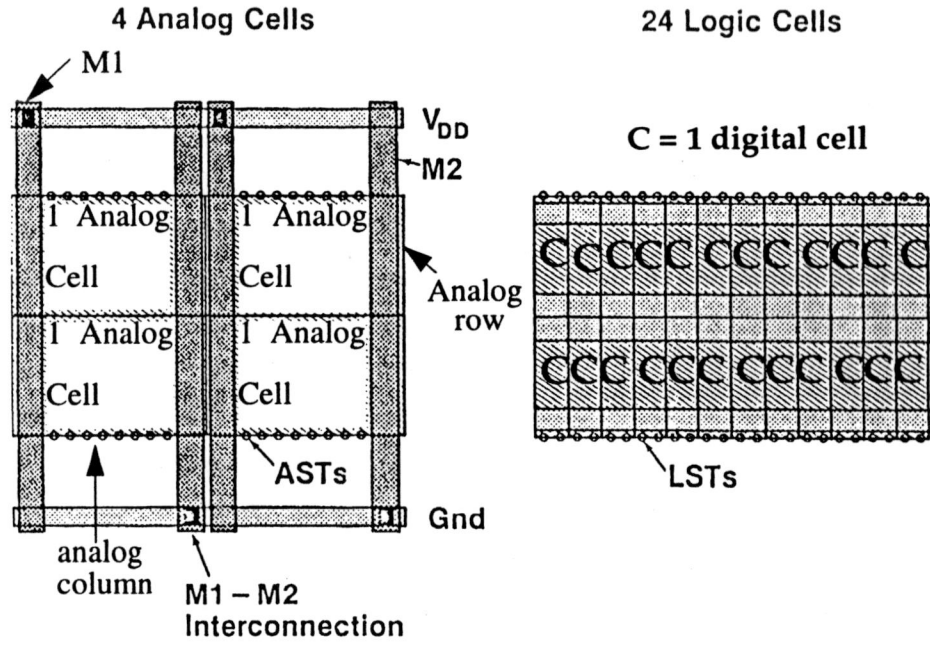

FIGURE 8.4 Simple power GRID (for IC with wiring channels)

Figure 8.5 shows a bus analysis for this power grid for 15 rows of 18 analog cells each. Part A of the figure shows the voltage dc distribution with 1ma of current being drawn in each of the analog cells. The power bus is connected to the chip pads for package wirebonding at the ends of the columns. Part B of the figure shows the 0.8 ohm dc bus resistance when 1ma is drawn in a worst case analog cell. A table at the bottom of Figure 8.5 compares the dc voltage drop and bus resistance of a grid versus tree bus connection scheme. The tree design has a bus drop of 149mv compared to 36mv for the grid. The tree design has a bus resistance of 5 ohms compared to the 0.8

Effect of Power Bus Structure on Noise coupling

ohms of the grid design. Note that in Part A that, even with a dc current drawn on the bus and a 0.8 ohm bus resistance, there is a large voltage variation across the power bus. Conventional single ended voltage biasing of circuit transistors current sources would result in large current variations across the bus due to this voltage gradient. When some of the currents entering the power bus are switching the voltage gradient across the bus will vary with time. If bias references are distributed as differential voltages or currents that are locally re-referenced to the local power rail the effect of this bias variation can be reduced.

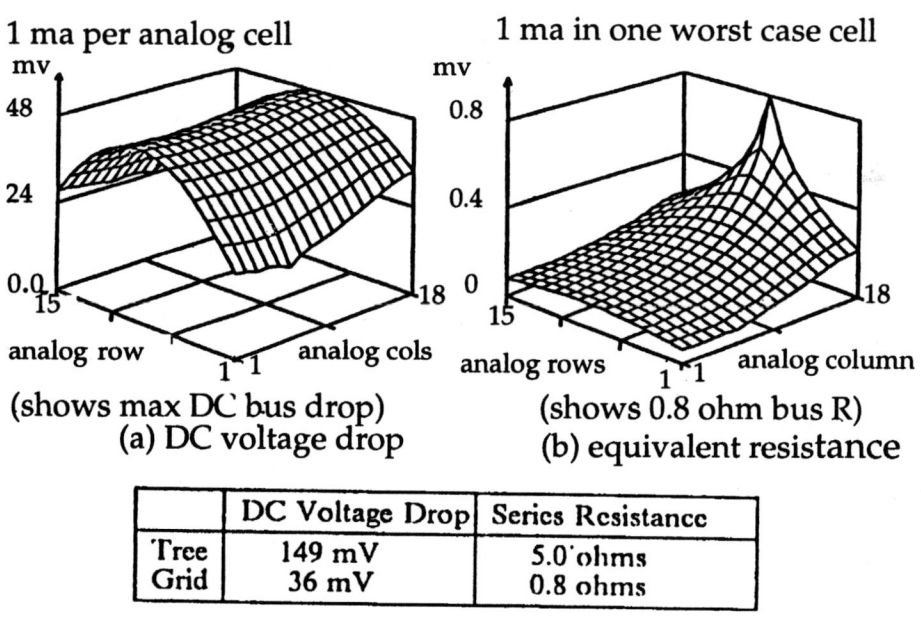

	DC Voltage Drop	Series Resistance
Tree	149 mV	5.0 ohms
Grid	36 mV	0.8 ohms

FIGURE 8.5 TREE TO GRID COMPARISON-bus electrical analysis.

Figure 8.6 shows a single rail grid power connection scheme where the switching and non-switching functions have been strictly partitioned to use separate chip power pads. This is usually required due to the large inductive bounce voltages determined previously. Figure 8.7 shows a similar situation except STAR instead of GRID power rail routing is used. It is not always possible to strictly partition switching from non-

Modeling Chip/Package Power Distribution   **153**

Modeling Chip/Package Power Distribution

switching functions and in this case the STAR power routing gives less coupling at the expense of a totally custom power bus. If some on-chip power supply decoupling capacitance can be provided the split-grid power bus scheme can give superior performance without a custom bus design. Reference [8.13] describes a fully automated power bus routing software system that routes the bus to achieve minimum bus resistance and coupling while minimizing chip area used by the bus. A performance function needs to be specified and the coupling mechanisms need to be quantified for the particular chip process. Clearly this is the approach needed to achieve an optimized power bus routing.

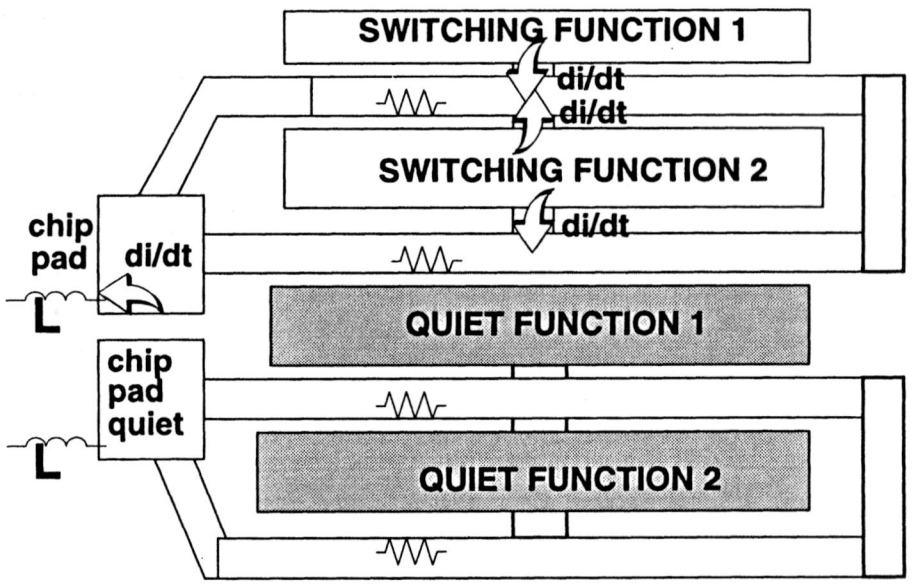

FIGURE 8.6 STAR power feed using Grids. (Best of Star and Grid)

One of the problems that arises with a separate power bus and pads for switching and non-switching functions is communicating signals between the multiple power domains. Because the separate power buses have different noise waveforms superimposed on them, there is a time varying difference between the bus voltages in different power domains. Circuits that drive voltages across power domains as well as circuits

Effect of Power Bus Structure on Noise coupling

that receive voltages from another power domain must contend with this time varying bus difference voltage. Figure 8.8 shows the situation where a CMOS signal driven from the quiet analog power domain, on the left, must be received by an inverter that resides in a noisy switching power domain. The finite rise time of the analog inverter output causes jitter in the received signal since the receiving inverter switching threshold is changing with time, relative to the received signal, due to its supply bounce. This is a common situation since often a clock is generated by an analog phase lock loop circuit and communicated to the switching power bus. The jitter can be reduced by decreasing the rise time of the analog output signal or reducing the voltage bounce on the switching supply. This problem often requires that clock generation circuitry reside in the same power domain as the circuitry that is driven by the clock. The problem can also be solved by passing differential signals across the power domain. This complicates the driver and receiver design since CMOS circuits are typically single ended in topology. Special differential receiver circuits are usually employed followed by a differential to single ended converter.

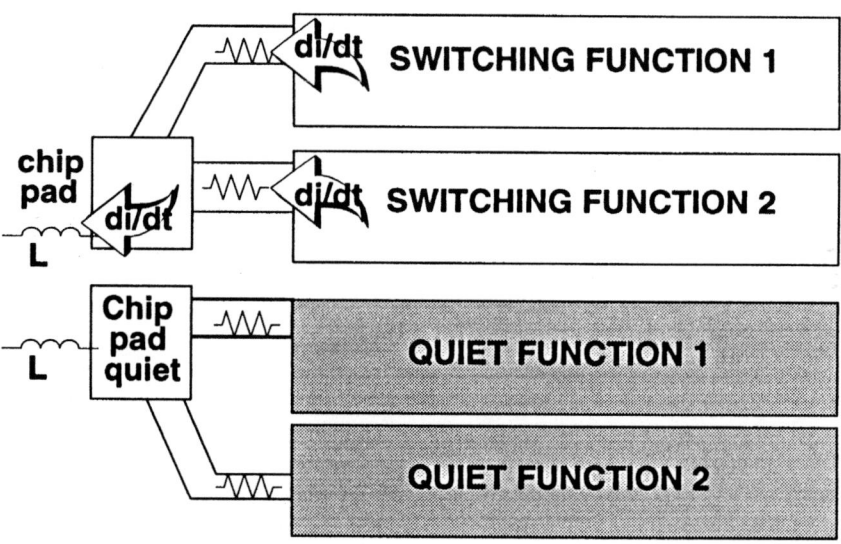

FIGURE 8.7 Split power feeds using Stars. (equal to a split grid but a custom bus).

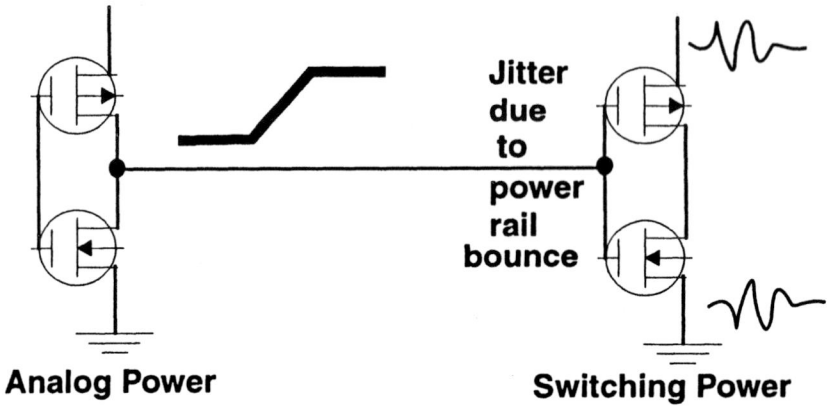

FIGURE 8.8 Communication between separate buses. (quiet to switching).

Figure 8.9 shows the situation in reverse where a signal is communicated from the switching power domain to the non-switching or analog power domain. The same jitter situation occurs and needs to be fixed in the same way: Reduce the power bounce and decrease the signal rise time. In this situation a differential driver or two out of phase single ended drivers can partially cancel the noise current coupled to the analog power rails when the digital driver switches. However, the noise coupled from the RLC digital noise is doubled by a differential or out of phase single ended driver scheme. Figure 8.10 shows a signal communicating between the switching power domain and an analog switch in the analog or non-switching domain. Note that the control signal to the analog switch carries the switching power supply bounce and resonance even when it is static or between switching events. The noise signal couples through the switch capacitance and enters the analog circuit. The signal coupling can be reduced by including one or more small series inverters in the analog power domain to remove the switching supply bounce signal. This situation is depicted in Figure 8.11. The goal is to keep the capacitance CA1 as small as possible to reduce the feedthru signal. The problem with this approach is that if there are many signals that need to be buffered in this way, there will be large amounts of switching current added to the quiet bus when these buffers switch. However, it is still usually preferable to couple noise into the supply rails instead of a sensitive circuit node. If the signal entering the analog power domain from a switching domain is a slow speed control signal, it is often possible to route the signal on a high resistance interconnect layer such as polysilicon. This added series resistance will further reduce the amount of

Effect of Power Bus Structure on Noise coupling

noise current coupled to the analog domain power rails. This added series resistance will also tend to isolate the two power domains. Note in Figure 8.11 that the capacitances Ca1, Ca2, Ca3, Cl1, Cl2, Cl3 and the channel resistances of the driver and receiver devices effectively connect the two power domains with a series RC circuit. The more signals that cross the two power domains the lower the R and the higher the C of the linkage. If a large number of wires cross the two domains the two power domains are effectively the same.

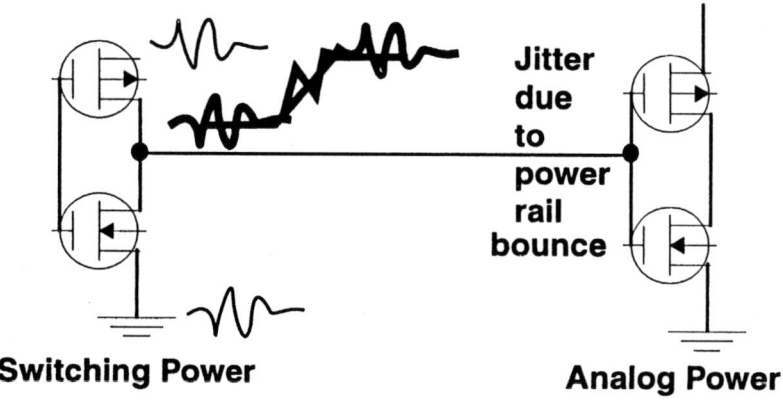

FIGURE 8.9 Driving logic to analog (noisy to quiet)-transient jitter.

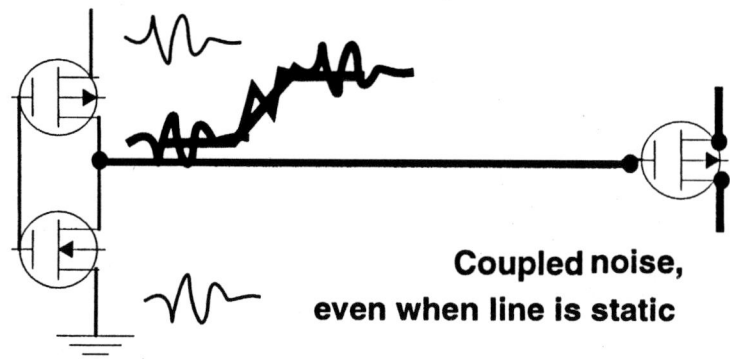

FIGURE 8.10 Driving logic to analog. (noisy to quiet)

Modeling Chip/Package Power Distribution

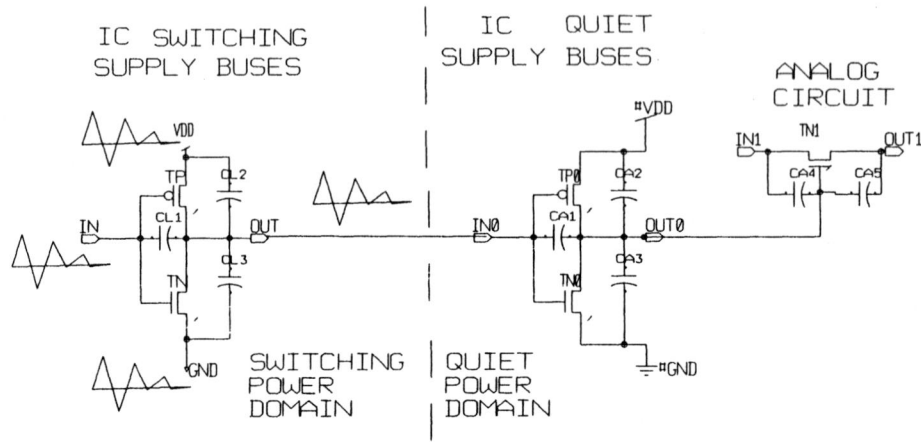

FIGURE 8.11 Buffering Logic Signals Entering Analog Terrain.

8.1.1 RLC CARD, PACKAGE AND CHIP RESONANCE

The power rails of an IC generally have the most capacitance to chip substrate and to each other due to the large number of devices connecting to them. This large capacitance together with the inductance and resistance from the chip, bonds, and package leads form multiple RLC resonant circuits with their own distinct natural frequencies. When stimulated by on-chip voltage and current transients, these RLC circuits will add their own characteristic noise frequencies. The sharpness of resonance and the resonant amplitude will depend on the Q of the system. The Q will vary with CMOS switching activity since the number of "on" gates or inverters partially determines the resistance or damping between the supplies. Tuning is usually required to prevent a power bus resonance from lining up with a system clock frequency. Tuning of the resonant frequencies is possible by controlling the amount of on chip capacitance and off chip inductance. Damping the RLC transient response can be accomplished by reducing the time constant $2L/R$ of the RLC circuit. This can be done by increasing the bus DC resistance or decreasing the system inductance. A variety of ways exist to increase on-chip capacitance but it is not easily reduced. The resonant frequency is usually raised and the damping increased by lowering the packaging inductance. This can be done with better packaging or increasing the number of package leads assigned to the power rails. Figure 8.12 shows a chip/package model for a mixed signal, 1.2u bicmos chip packaged in a 68 PLCC. This IC is reported in reference [8.2]. This IC used a 0.006 ohm-cm P+ bulk substrate (0.1 ohm per square sheet resistance) with a 10 ohm cm P- epitaxial layer (22Kohm per square sheet resistance.) The chip substrate is approximately 600 microns thick (a 5 inch diameter wafer) and the P- epitaxial layer is 6.5 microns thick. Because of the degenerative doping of the P+ bulk substrate, the substrate is considered a single node for modelling purposes [8.2]. This is a good assumption since the resistance across the lower P+ substrate for any square chip is only 0.1 ohms.

The chip was partitioned into an analog and digital portion. Figure 8.14 shows the floorplan of this IC. Figure 8.15 shows the photomicrograph of the IC. Figure 8.16 shows how the chip power pins were allocated. The chip, of this example has two distinct power supplies, an analog and a digital or CMOS supply. The Vdd rail was 5 volts and the lowest voltage is ground. Neither the digital or analog ground rails contain any substrate contacts. This accounts for why RLGS and RAGS resistance values are so high in Figure 8.12. There is a 14 micron wide metal contact tied to a 10 micron wide P+ diffusion (96 ohms per square sheet resistance) that wraps around the entire outside perimeter of the chip. This contact serves as the only substrate contact for the entire chip. Because of the spreading resistance associated with the surface channel stop P- diffusion (effectively 3.4Kohm/square sheet resistance), the effective contact width is increased by approximately 20 microns. When the metal contact is tied to a chip pad in a single spot (bottom center of the chip) it affords a 3 ohm connection to the P+ bulk. This is sufficient to prevent latch-up since the P+ bulk has a resistance of only 0.1 ohms for a square chip.

Because of the series resistance of the substrate metal connection (0.057 ohm/square) the contact is only carrying significant current for about 2 mm of length either side of the chip pad that is bonded out. If the two chip pads, tied to the substrate ring, would have been placed more than 4mm apart the connection resistance to substrate would have halved to 1.5 ohms. The single contact on the outside edge of the chip has advantages for latch-up prevention. Current injected into the p- epi layer, inside the chip, is collected by the P+ bulk and is conducted to the edge of the chip. The surface contact at the edge of the chip pulls the carriers back into the P- epitaxial layer for collection by the contact. This is done at the edge of the chip where there are no devices to be affected by the voltage gradient in the epi region due to carrier collection.

Modeling Chip/Package Power Distribution

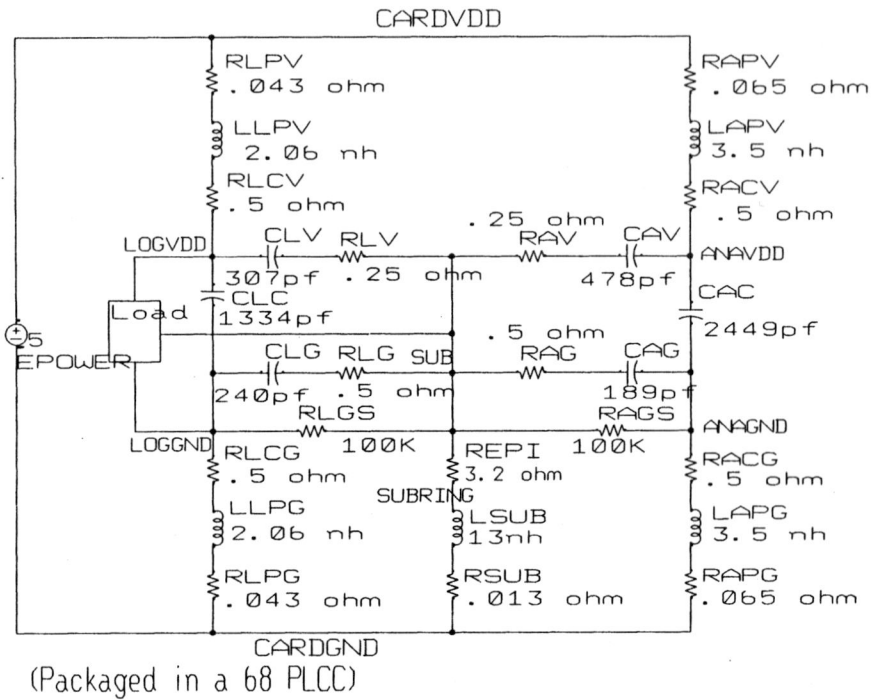

(Packaged in a 68 PLCC)

FIGURE 8.12 Chip/Package power model for P+ bulk with P- epitaxial substrate.

- Package Model Element Descriptions for Figure 8.12.

The second character in the element name signifies if the component is ANALOG (A) or LOGIC (L)

RLPV, RLPG, RAPV, RAPG- resistance of package power connections

LLPV, LLPG, LAPV, LAPG - inductance of package power connections

RLCV, RLCG, RACV, RACG, - resistance of chip power connections

CLV, CAV - capacitance from chip Vdd to substrate

CLG, CAG - capacitance from chip ground to substrate

RLV, RAV - resistance from NWELL (Vdd) to substrate

160   *Simulation Techniques and Solutions for Mixed-Signal Coupling in ICs*

Effect of Power Bus Structure on Noise coupling

RLG, RAG - resistance under the N+ diffusions to sub

RLGS, RAGS - substrate contacts that are connected to chip ground.

REPI - epitaxial layer resistance between the substrate contact and the P+ substrate.

LSUB - package inductance to tie substrate to card gnd

RSUB - package resistance to tie substrate to card gnd

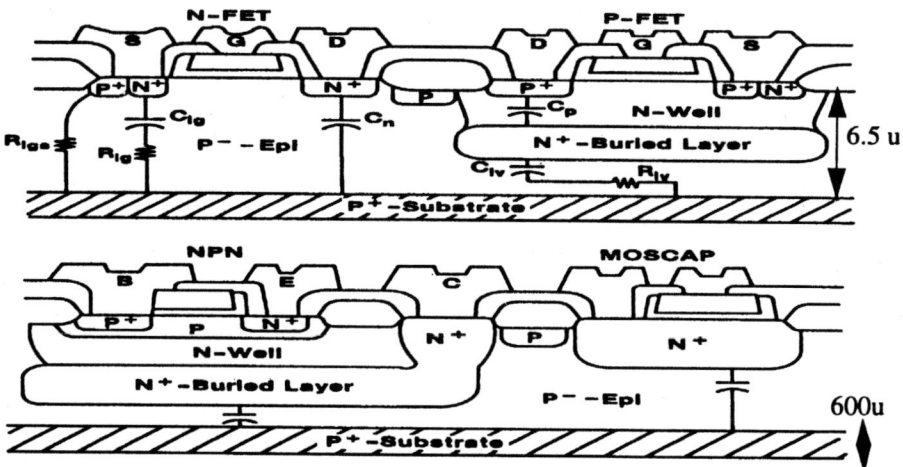

FIGURE 8.13 Cross section of a Bicmos process with parasitic elements.

- Cross-section Element Descriptions for Figure 8.13.

  Rlgs - Substrate contacts that are connected to ground potential somehow.

  Clg - Capacitance from ground to substrate. This is mostly N+ diffusions that are tied to ground. Ground wiring over thick capacitance contributes to Clg.

  Rlg - Resistance under the N+ diffusions to substrate.

  Cn - Capacitance from N+ source, drains to substrate.

  Cp - Capacitance from P+ source, drains to substrate.

  Clv, - capacitance from Vdd to substrate. This is mostly NWELL that is diffused into P-Minus Epi. Vdd wiring over thick oxide contributes to this capacitance.

  Rlv - Resistance under the NWELL down to the substrate.

Modeling Chip/Package Power Distribution

Modeling Chip/Package Power Distribution

Figure 8.13 shows a process cross section with parasitic elements overlaid. These parasitic elements combine to produce chip capacitance and resistance terms of Figure 8.12. The inductance and resistance values for the package in Figure 8.12 can be calculated. The chip inductance values can be ignored for purposes of determining the power rail model. The chip capacitance and resistance values must be determined and/or extracted from the various junction geometries and doping concentrations. For purely Cmos logic an estimate can be made using the total area of non-switching nwells and the total area of non-switching n+ source and drains in the p- epitaxial layer. Knowing this and the junction doping concentrations and the average amount of reverse bias will allow a rough calculated estimate. Alternatively, a detailed analysis and calculation can be made for each unique cell in a cmos logic library. By summing the occurrences of each cell a total value can be obtained. Note in Figure 8.12 that the resistances from the digital and analog ground rail to substrate (RLGS and RAGS) are set equal to 100Kohm because in this example there are no substrate contacts on the ground busses. There is a separate large contact that connects to the chip substrate and ties to a dedicated package bond and pin. This contact is depicted in Figure 8.14. It is clear that including substrate contacts on either or both ground rails will connect the bus to substrate via a few ohm resistor.

It is important in the model of Figure 8.12 that the capacitances associated with on chip switching nodes not be included directly in the model. The block labelled Load in Figure 8.12 should contain the capacitances associated with switching nodes. In Figure 8.13 the capacitances Cn and Cp from switching source and drains and the gate capacitances tied to those sources and drains is not included in Figure 8.12 but rather in Figure 8.19. The capacitances in Figure 8.12 are the capacitances that are not directly connected to switching nodes, and contribute to tuning the power supply RLC circuit. The block labelled Load in Figure 8.12 will contain all capacitances from switching nodes to supply rails and the substrate. This "output coupled noise" from switching nodes acts as the excitation that stimulates the power supply and substrate RLC circuits. Circuit block "Load" must also contain any estimates of wiring capacitance from switching nodes to chip substrate and the supply rails.

It is possible to measure the chip capacitances of Figure 8.12 on actual hardware. This must usually be done at zero volts applied to make sure no junctions are forward biased. The value measured at zero volts must be reduced by the appropriate amount of reverse bias that occurs at the normal 5 volt bias condition. For example the 3 logic unknown capacitors (CLC,CLV, and CLG) can be determined by measuring the capacitances with an RLC meter between logic Vdd, logic ground and chip substrate using a series and parallel configuration measurement and solving for the unknowns.

Effect of Power Bus Structure on Noise coupling

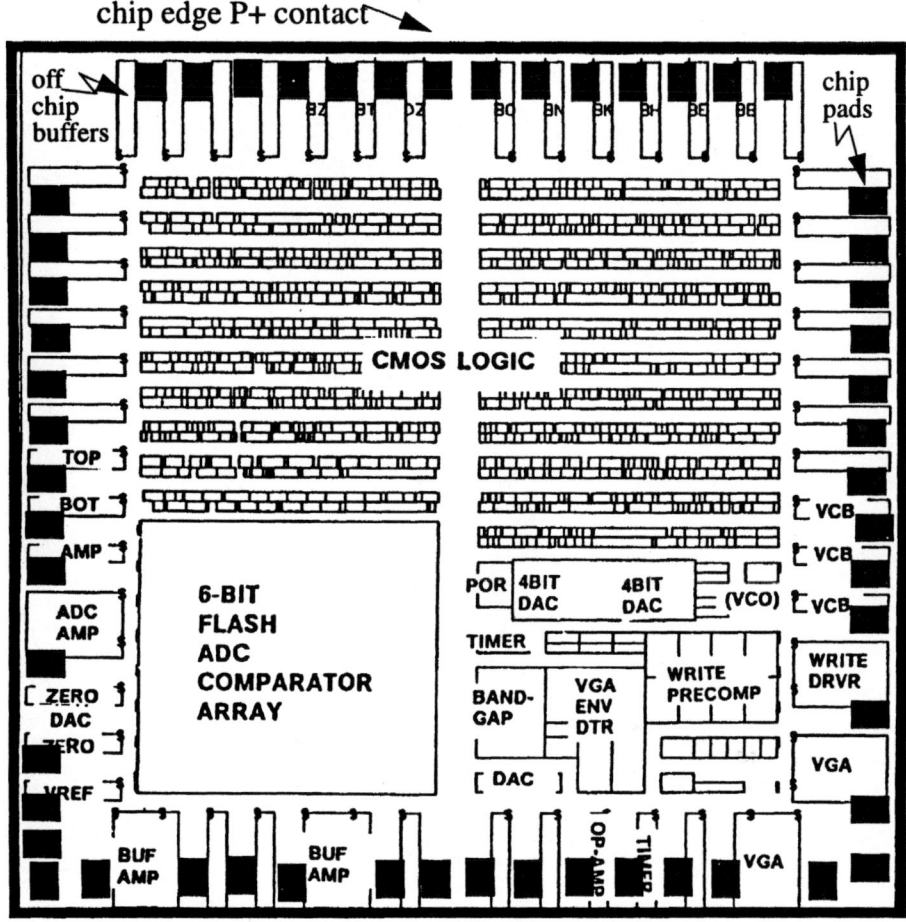

FIGURE 8.14 Chip example floorplan.

As stated earlier, for large chips with an inverting logic family like CMOS, there is, on the average, one rising signal for every falling signal. The total chip average of injected current into the substrate is small with the net affect that the substrate RLC circuit is stimulated. Figure 8.20 shows a scope picture of a chip ground supply rail resonating with its own natural frequency after being stimulated by a clock edge. The IC has all of its output drivers disabled for this picture so that the power rail has a balanced excitation from only on-chip CMOS logic switching. The resonant frequency is approximately 72Mhz in this example.

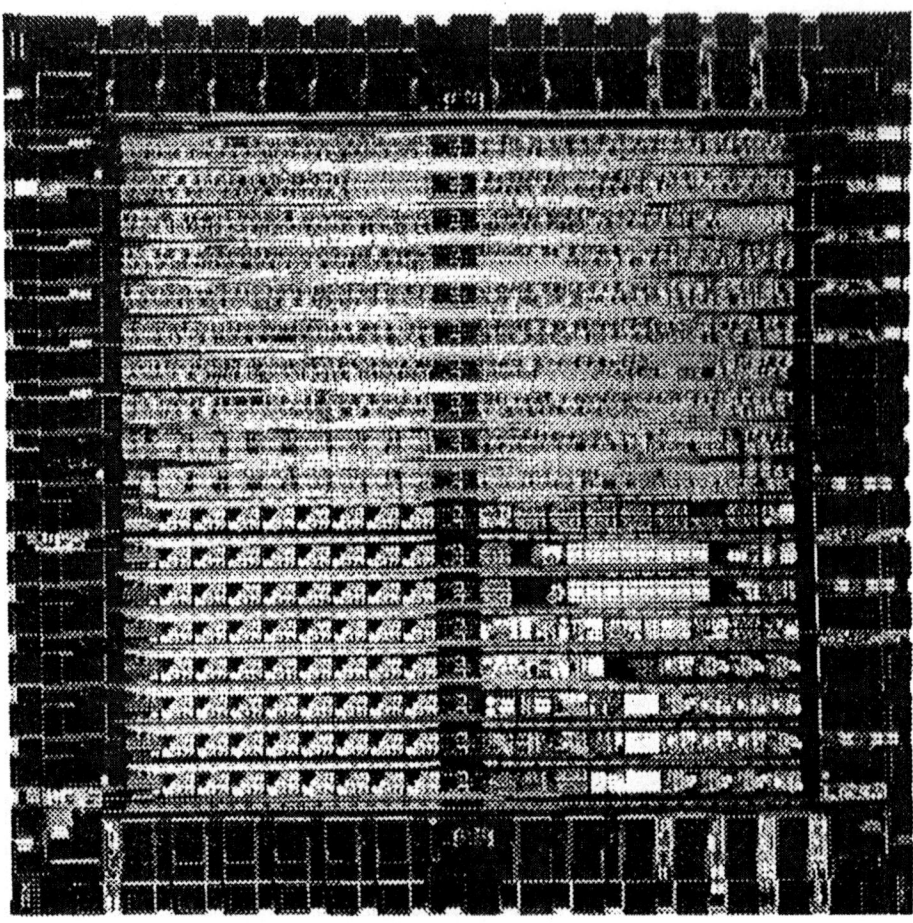

FIGURE 8.15 Example Chip Photomicrograph

Using the model of Figure 8.12 the resonant frequency and amplitude of the signal on the power rails and chip substrate can be predicted. The excitation of the switching

Effect of Power Bus Structure on Noise coupling

logic is approximated by the circuit of Figure 8.17. Figure 8.17 shows two strings of seven inverters. The bottom inverter string is driven by a signal 180 degrees out of phase relative to the top inverter string. If each of the inverters is the same size then the amount of noise coupled into the substrate from output nodes switching will be zero. The net effect will be the excitation of the substrate RLC resonance. Figure 8.19 shows the circuit used for each inverter of Figure 8.17. The inverter model includes the capacitances from the switching nodes to the chip substrate and chip power supply rails. Each inverter model includes a multiplication factor to increase the amount of output switching power without modeling large numbers of inverters. Each inverter model of Figure 8.17 can have a different multiplication factor.

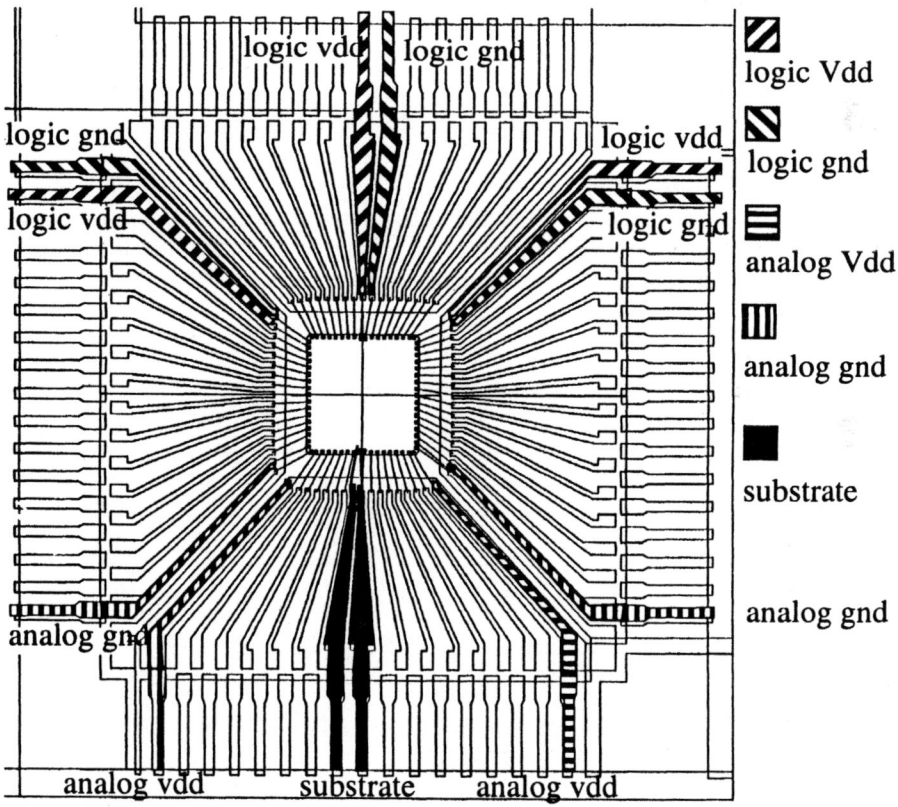

FIGURE 8.16 68 PLCC Leadframe drawing with Power I/O Assignment.

Modeling Chip/Package Power Distribution

There are seven inverters in each string in Figure 8.17 to take into account the fact that the output switching occurs over a number of gate delays after the clock edge. This number is valid for this example only. Some designs will require switching over a longer or shorter time period. Since substrate coupled noise power or energy is directly proportional to the amount of switching power or energy, it is important that the combined, multiplied inverter strings of Figure 8.17 dissipate the same switching power as the actual IC will. If the IC has a number of different operating modes with different switching power and different clock delays for the switching power then a different load model of Figure 8.17 will be required for each. Clearly, the model of Figure 8.17 will allow the effect of the noise coupled to the chip substrate from the switching logic circuits to be seen on the analog power rails. This noise will be in the form of both resonant frequency components and node switching frequency components.

If the analog circuit is connected to the analog supply and substrate terminals of Figure 8.12 while the logic Load is switching the effect on the analog circuit can be simulated. The correct distribution of the coupled frequency components will also be present. Likewise a digital circuit can be connected to the digitals supply and substrate terminals of Figure 8.12 while the digital Load is switching and the affect of the supply bounce can be seen on the digital circuit. The bus to bus coupling and jitter problems will also be manifiested by this simple model.

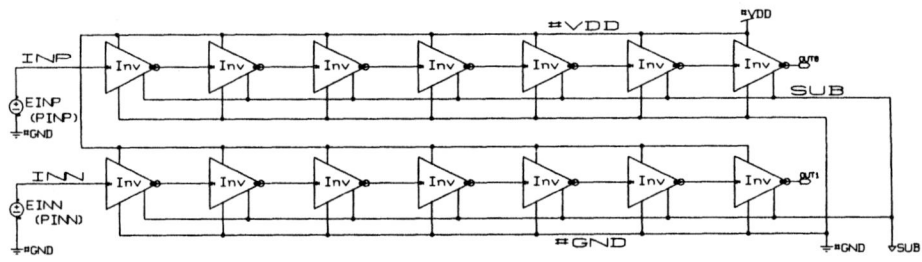

FIGURE 8.17 Approximation of CMOS logic switching activity [8.14].

Effect of Power Bus Structure on Noise coupling

The analog circuit can then be optimized to give maximum rejection in the frequencies of resonance and coupled noise and the digital circuit can be optimized to tolerate the environment as well. The model also allows for the resonances to be tuned away from problem frequency ranges by tailoring the chip capacitances and the number of package pins assigned each power supply. It is also possible using the model of Figure 8.12 to determine the optimum distribution of allotted package pins to each power rail and substrate to achieve the minimum coupled noise. Without modelling it is difficult to determine how the package pins should be distributed between the analog supplies, logic supplies, and substrate ties. Care must be taken that the supply capacitances of the analog or digital circuit being simulated are subtracted from the whole chip model of Figure 8.12.

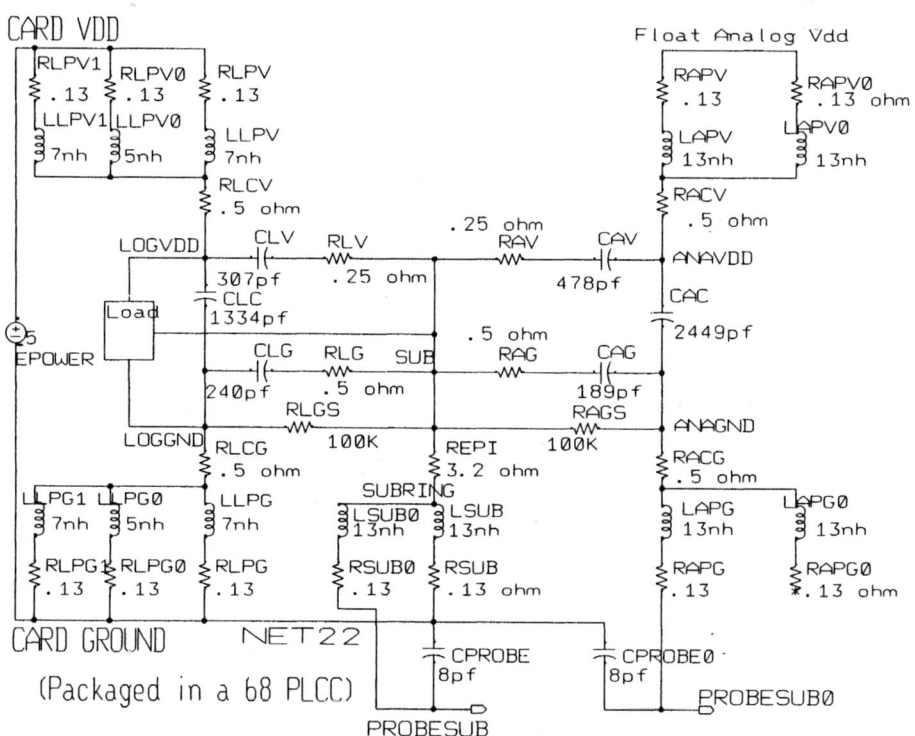

FIGURE 8.18 Model showing individual pins and probing capacitance.[8.14]

Modeling Chip/Package Power Distribution

Modeling Chip/Package Power Distribution

Figure 8.21 shows hardware to model correlation of the model of Figure 8.12 in predicting amounts of chip noise on the chip substrate. To make this measurement of the voltage at node SUB in the model of Figure 8.12, the analog supplies were lifted from the card and floated. One of the analog ground pins was used as a probe into the internal SUB node. Only measurement current flows through the 8PF probe plus parasitic capacitance and the bond inductance. A Tektronix 6201, 3PF, active probe with 900MHZ of bandwidth was used. One of the two substrate ring pins could have been used to probe node SUBRING of Figure 8.12 but that would have included the drop across the REPI resistor. Figure 8.18 shows the individual pins for the chip package power model. The mutual inductances have been used to reduce the self inductances for the switching supplies. One of the substrate ring pins is always available to probe the substrate ring. A floating analog ground pin is used for a probe into the actual substrate node SUB in the model.

For the top trace of Figure 8.21 the multiplication factors used on the inverters from left to right (Figure 8.17) were top string:22,20,20,20,1,1,1 and bottom string: 19,20,20,20,1,1,1. This is a basically balanced excitation and the figure reflects mainly the RLC resonant damped sinusoid. The high frequency sinusoid is due to the LC of the probing measurement system. For the bottom trace of Figure 8.17 the multiplication factors used on the inverters from left to right (Figure 8.17) were top string:10,10,10,10,1,1,1 and bottom string: 30,30,30,30,1,1,1. This is a a very unbalanced excitation and the output waveform reflects the direct output coupling from the bottom string along with the damped power supply resonance that follows.

The inverter weighting factors for this example were determined by trial and error but they can be predicted without hardware by simulating the logic circuits involved to determine switching factors and the amount of balanced or un-balanced switching that occurs. Note that there is no guarantee that the switching activity will be constant from clock edge to clock edge so these weighting factors can even change from clock edge to clock edge. This approximation technique is not as accurate as simulating with the actual circuits but it can give a surprisingly close approximation with very little simulation time. Very large circuits must be either partitioned many times for complete simulation of each partitioned or simulated with a lumped model equivalent as presented here.

These particular photographs were taken with the clock frequency slowed down enough so that the transients could die down completely between clock edges. If the clock frequency is increased enough then the transients do not die out and the waveforms from the previous clock edge interact with the waveforms from the next clock edge. The constructive and destructive interference that occurs radically changes the

Effect of Power Bus Structure on Noise coupling

supply and substrate waveforms. The resistance in the circuit will damp out the sinusoid with a time constant of $2L/R$.

The model of Figure 8.17 is a simplification for CMOS logic and is fairly accurate for generating the proper spectrum of coupled frequencies and exciting resonances in the power supply model. More accuracy could be obtained by including static or dynamic latches in the model as well as BiCMOS circuits and other switching circuit types as the particular IC design required. Additionally, the weights on the switching stages could change with time to emulate variations in switching activity. Behavior models of each logic function could also be modified to include the parasitic elements to inject the proper noise signals into the supplies and chip substrate. This would allow for simulating the entire chip inside the package model to get near exact stimulus and coupled noise levels. The simple model of Figure 8.12 can be modfied for the particular package and can include die bonds from chip or leadframe pins to the chip die paddle etc.

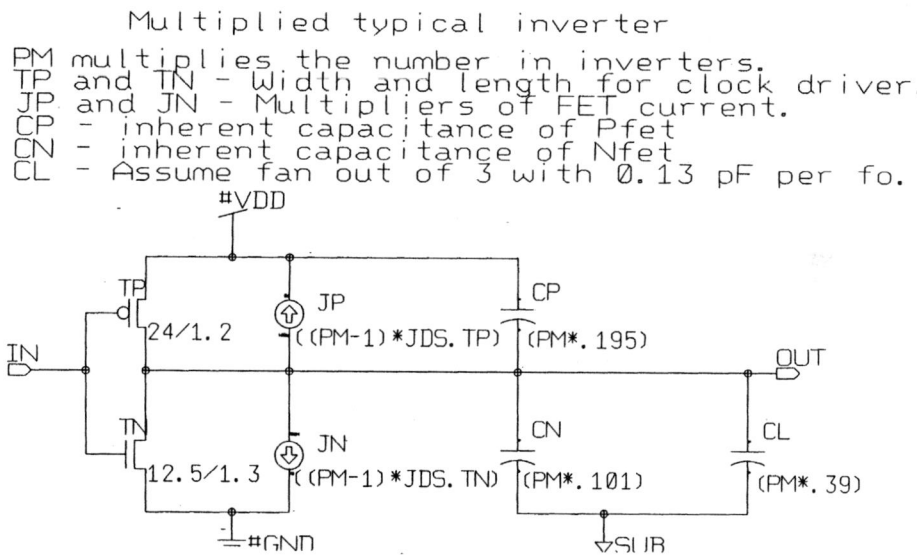

FIGURE 8.19 Multiplied inverter model used to produce logic stimulus.

Modeling Chip/Package Power Distribution

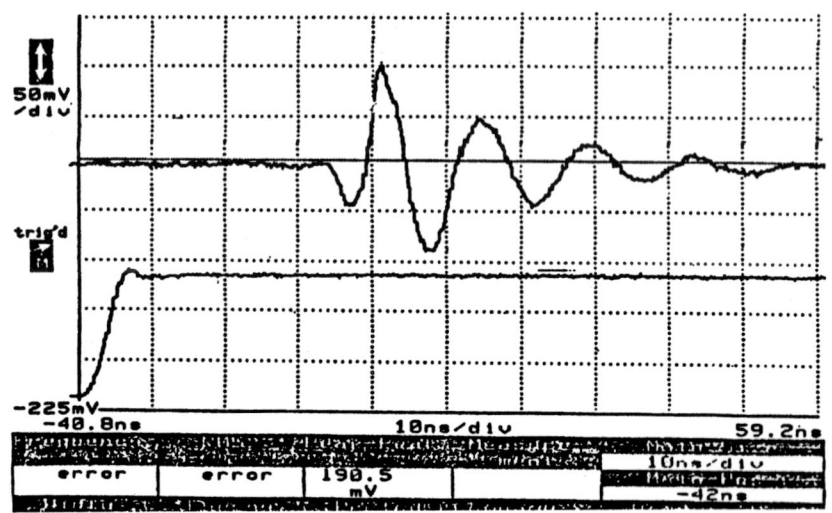

FIGURE 8.20 Power supply resonance when stimulated by a clock.

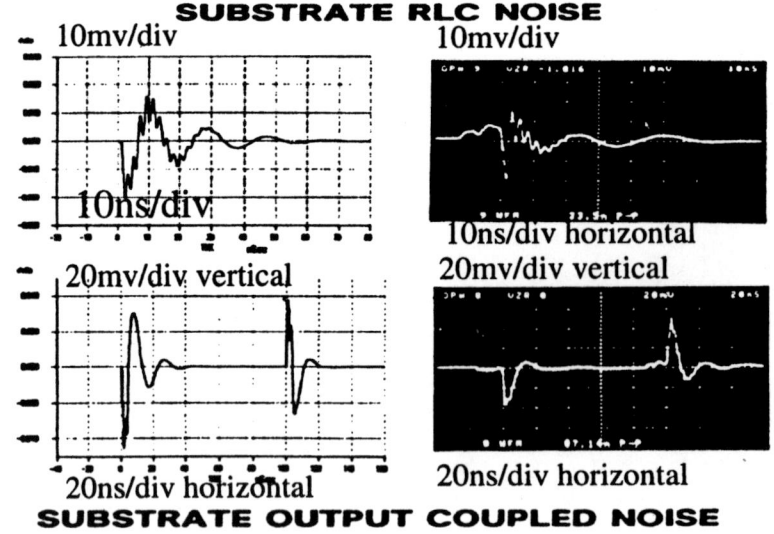

FIGURE 8.21 Power RLC ring noise versus I/O switching noise.

Effect of Power Bus Structure on Noise coupling

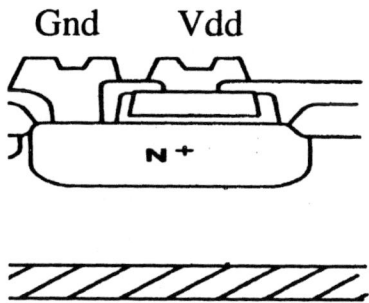

FIGURE 8.22 Connection of MOSCAP to get more capacitance to substrate from Gnd.

8.1.2 Cancelling CMOS RLC Coupling to Substrate

In cmos technologies where the ground and Vdd rails bounce out of phase with one another, it is possible to first order cancel the capacitive current injected to the substrate by making equal the capacitance from ground and Vdd to substrate. This is most useful in a P+ bulk/P- epitaxial processes where the substrate can be tied to a separate non power carrying contact or analog ground, removing the requirement that the switching ground bus be resistively tied to substrate. With a P+ bulk/P- epi process the Vdd to substrate capacitance is usually largest due to the large nwell to substrate area. The addition of thin oxide and/or junction capacitance between switching ground and substrate is required to balance the capacitance. Figure 8.22 shows a connection configuration for a MOSCAP that adds supply decoupling and adds a greater capacitance from ground to substrate. It does this by tying Vdd to the poly gate and ground to the N+ backplate of the MOSCAP. Since the substrate is also at ground potential this PN junction is biased at zero volts to give maximum capacitance to substrate. Figure 8.23 shows the logic Vdd and ground rails ringing out of phase from the model of Figure 8.12 on the left and from actual hardware on the right. This picture was also taken with the analog supplies floating so that their resonance and loading effects would not be seen. If there is not perfect cancellation the substrate waveform will look like a reduced amplitude version of the rail with the larger capacitance to substrate. If substrate contacts are placed on the logic ground bus then the substrate waveform will almost always reflect the logic ground rail phase.

Modeling Chip/Package Power Distribution

Modeling Chip/Package Power Distribution

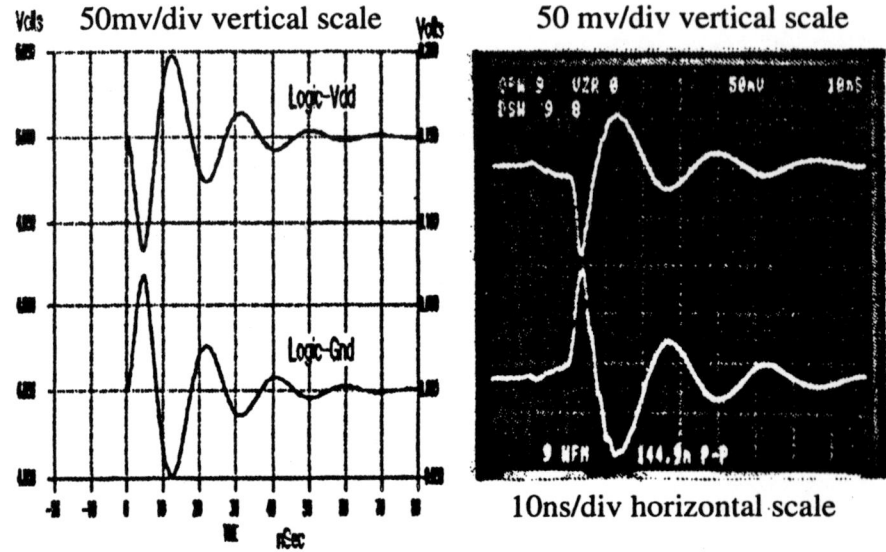

FIGURE 8.23 Cmos ground and Vdd ringing out of phase.

Figure 8.24 shows a simulation of the noise on chip substrate using the model of Figure 8.12. As the capacitance from logic ground to substrate, CLG, becomes closer to the capacitance from logic Vdd to substrate, CLV, in value the noise waveform on the substrate drops dramatically. This simulation was done for the same inverter weighting values as in the top trace of Figure 8.23 for a balanced substrate excitation. Output coupled noise due to an unbalanced excitation will not be cancelled since it is locally coupled directly to the substrate. Its cancellation is possible if locally there is always, for every signal, a matching 180 degree out of phase signal to cancel the substrate injection.

Figure 8.25 shows the simulated affect that added, on-chip, digital supply decoupling capacitance has on chip substrate noise. Note that as the decoupling capacitance, CLC, from the model of Figure 8.19 increases the resonant frequency of the substrate noise drops proportionally with the inverse square root of the capacitance as expected. Notice that the initial output switching noise capacitively coupled directly to the substrate for time less than 7 ns is independent of the decoupling capacitance. This is because the amount of current coupled directly to the substrate is dependent on the voltage swing at each output node, the rise or fall time of the transient, and the capacitance to substrate. It is not strongly dependent on the decoupling capacitance. Also note that the area under any of the curves is equal. This is a statement that the coupled energy is equal, depending mainly on the input switching power, and the decoupling capacitor only tailors the frequency of coupling.

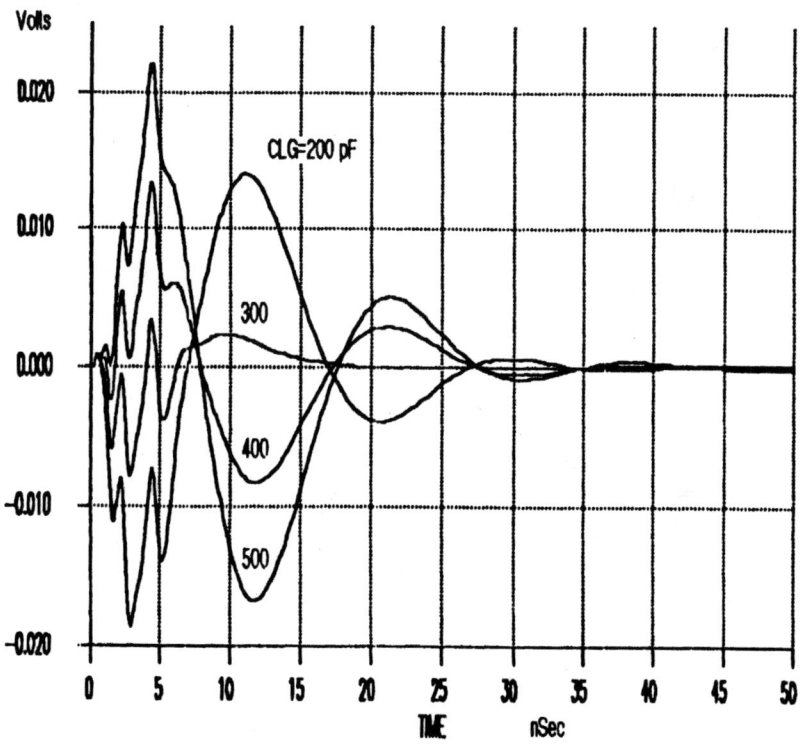

FIGURE 8.24 Substrate Noise As Logic Vdd and Gnd Capacitance is balanced.

Modeling Chip/Package Power Distribution

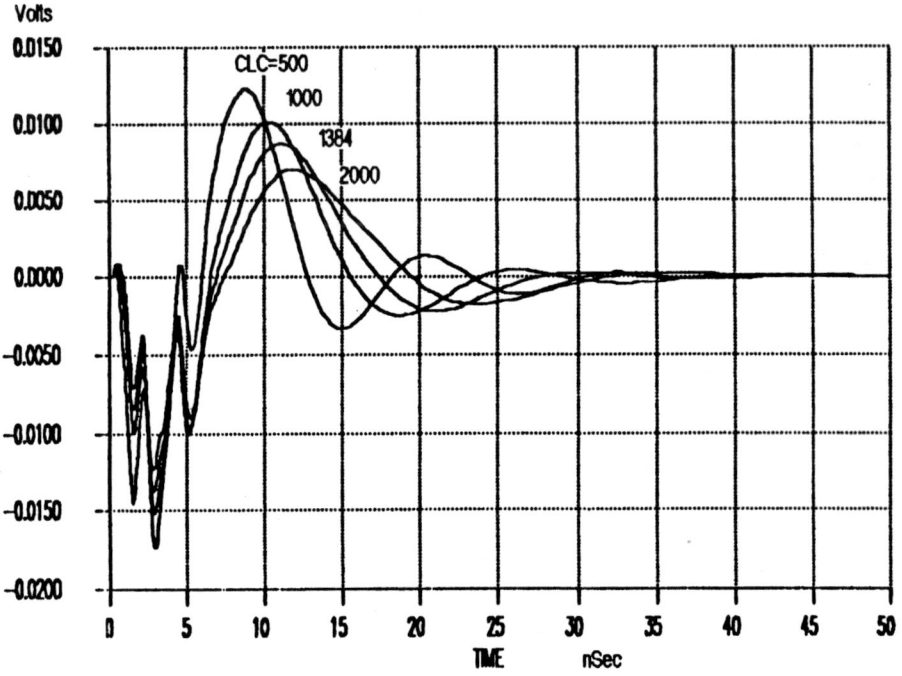

FIGURE 8.25 Effect of Switching supply decoupling on Substrate noise.

If off chip referencing of chip signals is not a concern the decoupling capacitance allows the chip circuitry to tolerate the large power supply inductive bounce by making it a common mode signal relative to both supply rails. The problem of communicating between power domains is still a problem, however. For some applications adding more on-chip decoupling capacitance can be detrimental if it lowers a supply resonant frequency to be near a signal frequency of interest. For other applications it can be detrimental to remove on-chip decoupling capacitance if it raises the supply resonant frequency to be outside the bandwidth of a common mode rejection loop.

Effect of Power Bus Structure on Noise coupling

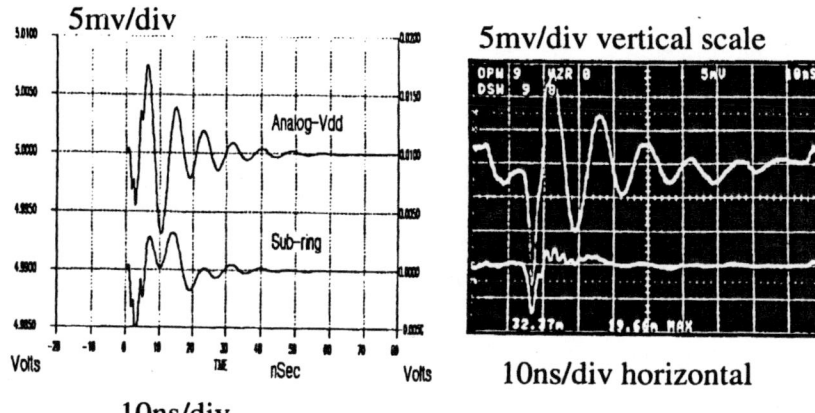

NOISE ON ANALOG VDD AND SUBSTRATE RING

FIGURE 8.26 Cmos power resonance coupled to substrate and analog power rails.

For Figure 8.26 the analog supplies were reconnected to the card and one pin was lifted from analog Vdd to act as a probe into the respective bus. In this case the now connected analog supplies act as a load to the substrate and influence the waveforms there. It is not possible to directly observe the node SUB but the node SUBRING in Figure 8.12 can be observed by lifting one of the two pins hooked to the substrate chip ring. Figure 8.27 shows this measurement configuration. Note that analog Vdd is ringing at about 100Mhz which is significantly higher than the frequency of the oscillation of the logic power RLC. The analog power bus has higher inductance (one pin was lifted to make the measurement), resulting in less damping and a longer time constant resulting from 2L/R. Thus it will ring longer once stimulated by a switching transient. Note that there is no circuit switching occurring on the analog power bus. The switching events on the logic power rails couple to the chip substrate and induce the transient and excite the resonance on the analog bus. Note also that the amplitude of the noise voltage on the analog Vdd is larger than the stimulating source from the substrate.

Modeling Chip/Package Power Distribution

Modeling Chip/Package Power Distribution

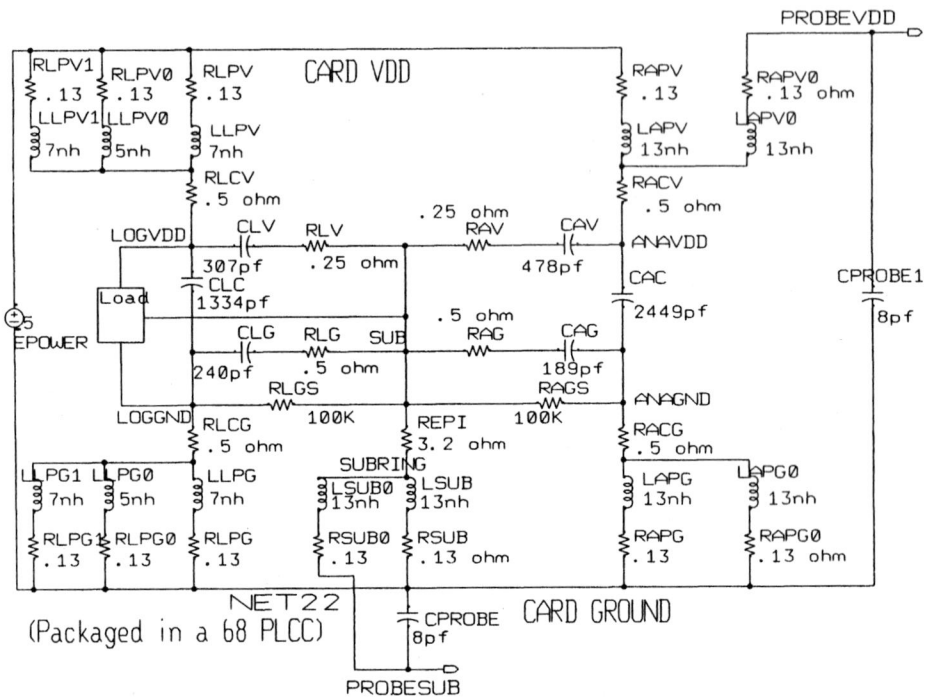

FIGURE 8.27 Measurement configuration to measure supplies and SUBRING.

The analog ground ringing waveform has the same phase as the analog Vdd waveform. This is primarily because the on-chip decoupling capacitor CAC in Figure 8.12 is keeping the signal difference between the analog supplies fairly constant. This capacitance also has the effect of putting the analog Vdd and ground package inductances virtually in parallel to reduce the effective inductance and raise the resonant frequency. Because of the smaller capacitance from analog ground to substrate versus analog Vdd the amplitude of the analog ground waveform is somewhat smaller. The waveforms on the logic ground and Vdd bus can also be observed without lifting a supply lead by using a CMOS inverter output driver pin that is static high (to measure Vdd) or static low (to measure ground). If the driver channel resistance is made small enough the probe LC connection does not induce significant error in the measurement of the bus waveforms.

8.1.3 OFF-CHIP DRIVER RLC COUPLING

The IC off-chip drivers also form an RLC resonant circuit which can be dominated by both off-chip capacitance and package inductance. This resonance is generally at a much higher frequency than power supply resonances and so coupling is greater through its capacitive connections to the substrate. Figure 8.30 shows pictorially the situation for an off chip driver. The off chip driver, shown as a simple inverter, has parasitic capacitance to both supply rails and the chip substrate. The output off-chip load is usually capacitive with its associated series inductances. The output voltage transient will ring with the characteristic frequency $f = \dfrac{1}{2\pi\sqrt{LC}}$ where there can be a different frequency for the Vdd return loop from the ground return loop. If a large on-chip decoupling capacitance exists from Vdd to ground the frequencies will be the same since the capacitance puts the loops in parallel.

Because the capacitances involved in the output loop are usually much smaller than in the supply loops the resonant frequency will be much higher. This resonant voltage ring on the output couples through CVDD, CGND, and CSUB to add to the noise integrated on the supply rails and substrate. The resonant voltage ring or any reflection transients on the output will also couple through ESD protect diodes D0 and D1 if ringing above the Vdd or below the ground rails transiently occurs. This ring is heavily dependent on the load and the rise and fall time of the driver. Controlled rise time output drivers help to control the typically 4 to 1 variation in standard CMOS driver rise and fall times. On low performance drivers a series resistance can be inserted to damp the ring. High Performance driver design requirements usually don't allow much tuning of the I/O driver output RLC circuit but placing large drivers close to their power return leads limits the inductive loop area for reduced radiated emissions as well as best performance. The use of balanced current steering drivers or those with controlled rise time and reduced voltage swing together with staggering switching of drivers on a bus limit or partially cancel the di/dt and dv/dt to the substrate and chip power rails for lower coupled noise amplitudes.

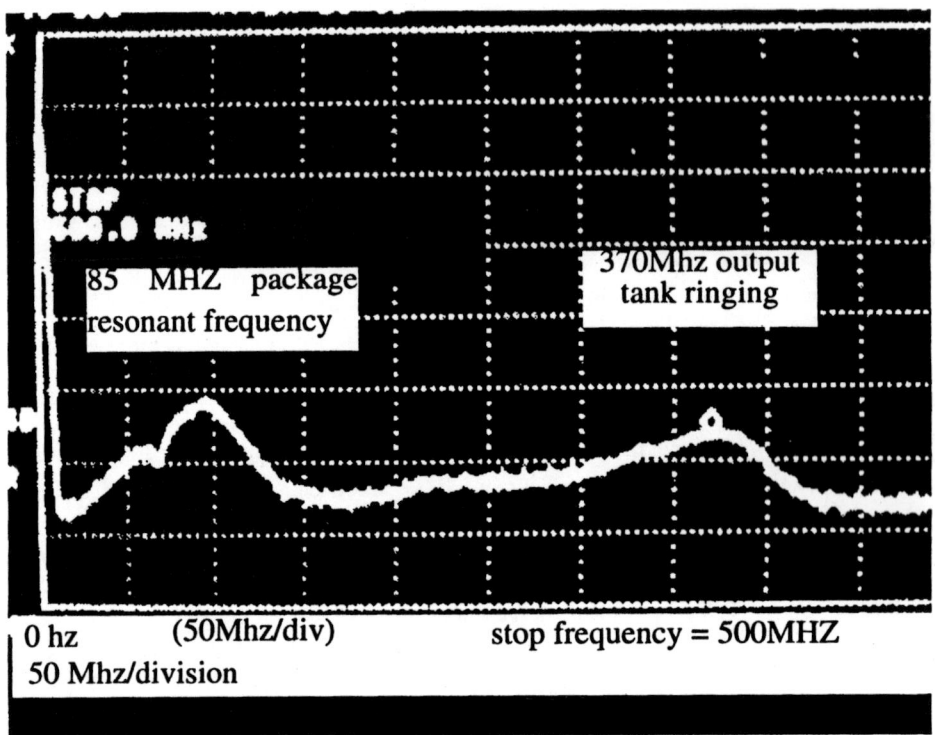

FIGURE 8.28 Substrate Coupled Frequency Spectrum with 1Mhz Clock.

Figure 8.28 shows the frequency spectrum on the chip substrate when both internal CMOS circuits and output drivers are allowed to switch. The clock frequency is 1Mhz which is sufficient to allow all transients to die out between clock edges. The 1Mhz clock is obscured in the spike near dc on the photograph. A frequency cluster occurs centered at 85Mhz which contains both the switching power domain and non-switching power domain package resonant frequency. A second frequency cluster occurs centered at 370Mhz which is the measured resonant frequency of the output driver RLC tank network. Some mixing is occurring which broadens the width of the frequency clusters. This mixing is lessened by the long clock period because at the start of the clock switching the supply rails are starting from a settled out condition.

Effect of Power Bus Structure on Noise coupling

Figure 8.29 shows the same substrate frequency spectrum when the clock frequency is increased to 27Mhz. This puts the third harmonic of the clock right near both the switching domain and non-switching domain package package resonant frequencies. Also, the supply transients cannot die out between clock cycles so constructive and destructive waveform interference occurs. Since the value of the switching CMOS currents depends on the supply voltage and the supply voltage is being modulated by the currents intermodulation products are created. Both Figure 8.28 and Figure 8.29 use the same reference levels and 10db/div amplitude scales. Peak spectrum amplitudes have increase roughly 20db or 10 times between Figure 8.28 and Figure 8.29 and their is no null in the spectrum anymore between the package resonant frequency and the I/O driver load resonant frequency.

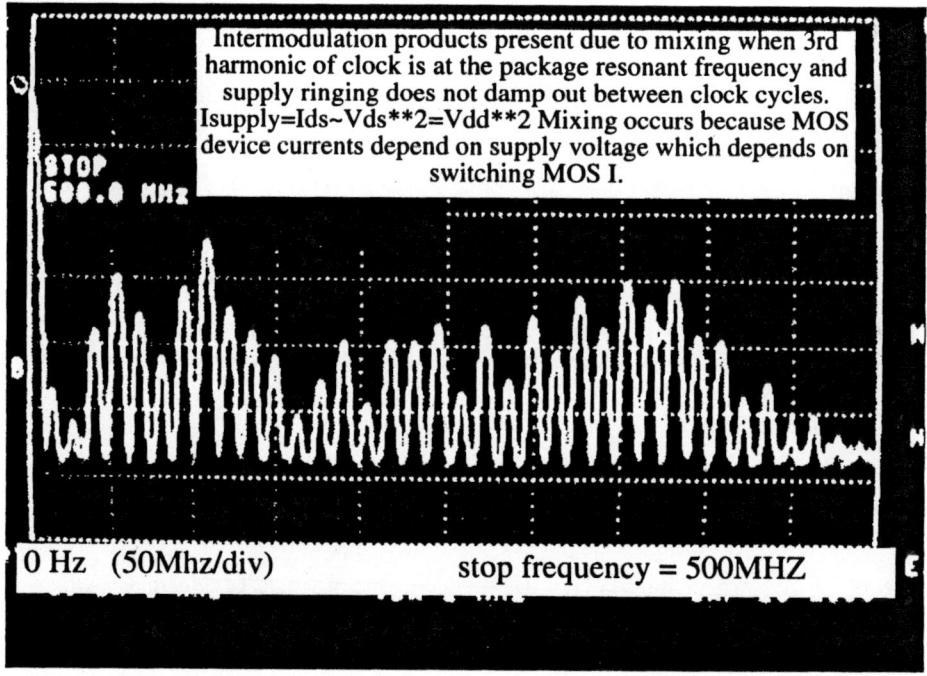

FIGURE 8.29 Substrate Coupled Frequency Spectrum with 27Mhz Clock.

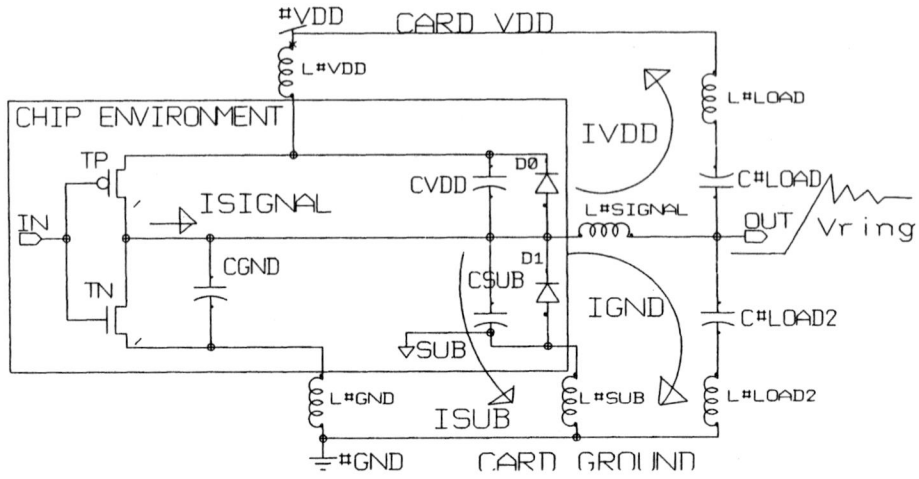

FIGURE 8.30 Output driver RLC tank circuit.

8.2 Summary

This chapter looked at power distribution systems in a packaged chip. A model for the chip/package system was developed and used to predict noise coupled through the chip substrate as well as power system resonance effects. Experimental data was given to compare to simulation and corroborate the model.

REFERENCES

[8.1] Timothy Schmerbeck, "Mechanisms and Effects of Noise Coupling in Mixed Signal ICs," EPFL, Switzerland course presentation, June 29-July 10, 1992.

[8.2] T. Schmerbeck, et al., "A 27 MHZ Mixed-Signal Magnetic Recording Channel DSP Using PRML," *Technical Digest of IEEE ISSCC*, pp. 136-137, Feb. 1991.

Summary

[8.3] D.K. Su, M.J. Loinaz, S. Masui and B.A. Wooley, "Experimental Results and Modeling Techniques for Substrate Noise in Mixed-Signal Integrated Circuits," *IEEE Journal of Solid State Circuits*, Vol. 28, No.4, pp. 420-430, April 1993.

[8.4] L. D. Smith, *et al.*, "A CMOS-Based Analog Standard Cell Product Family," *IEEE Journal of Solid-State Circuits*, Vol. 24, No. 2, pp. 370-379, April 1989.

[8.5] L. D. Smith, T. Schmerbeck, et al., "A CMOS-Based Analog Standard Cell, *IEEE Journal of Solid-State Circuits*, vol. 24, No. 2, pp. April 1989.

[8.6] Notes for the IEEE SSCTC workshop on NOISE in MIXED A/D ICs, held in Williamsburg Va, Sept 6-7, 1990. Tim Schmerbeck and Larry Smith. There were no proceedings,

[8.7] T. Gabara, "Reduced Ground Bounce and Improved Latch-Up Through Substrate Conduction," *IEEE Journal of Solid-State Circuits*, 23, No. 5, pp. 1224-1232, October 1988.

[8.8] Henry W. Ott, "High Speed Digital Design," Electromagnetic Compatibility Seminar, Henry Ott Consultants, January 1992.

[8.9] Dr. Tom Van Doren, "Grounding and Shielding Electronic Systems," NTU Satellite Network, February 1991.

[8.10] R. Philpott, T. Schmerbeck, et al., "A 7Mbyte/sec (65MHZ), Mixed Signal Magnetic Recording Channel DSP Using Partial Response Signalling with Maximum Likelihood Detection," *IEEE Custom Integrated Circuits Conference*, May 1993.

[8.11] N. Verghese, D. Allstot, and S. Masui, "Rapid Simulation of Substrate Coupling Effects in Mixed-mode IC's," In *Proceedings IEEE Custom Integrated Circuits Conference*, pp. 18.3.1-18.3.4, May 1993.

[8.12] T. Schmerbeck, "Design Strategies for Reducing the Effects of Noise Coupling in Analog and Mixed-Mode ICs," in presentation for course on *Practical Aspects of Analog and Mixed-Mode IC Design*, Beaverton Oregon, May 18, 1993.

[8.13] B. R. Stanisic, R. A. Rutenbar, and L. R. Carley, "Power Distribution Synthesis for Analog and Mixed-Signal ASICs in RAIL," In *Proceedings IEEE Custom Integrated Circuits Conference,* pp. 17.4.1 - 17.4.5, May 1993.

[8.14] Internal IBM Technical Report by Larry Smith, IBM, San Jose, CA.

[8.15] Kazuo Kato, Hideo Sato, Yasuji Kamata, Kenkichi Yamashita, and Seiichi Ueda, "A 300-MHz Monolithic Video Current Driver for High-Resolution CRT Applications," *IEEE Journal of Solid State Circuits,* Vol. 24, No.4, pp. 1110-1117, August 1989.

CHAPTER 9

Controlling Substrate Coupling in Heavily-Doped Bulk Processes

The chip substrate acts as a collector and distributor of noise on the IC. The amount of coupling varies depending on the structure and doping of the chip substrate as well as how it is tied to its assigned voltage potential on chip. Essentially every chip voltage transient on chip signal wires, I/O pads, and power rails is capacitively coupled to the chip substrate. This includes energy from card reflections back to I/O pads and transmitted to the substrate via I/O protect devices. Experiments done at IBM and outside have determined that the noise energy coupled to and from the substrate is proportional to the total chip switching power as well as to the logic power rail inductance and chip substrate tie inductance. Coupled noise peak voltages are frequency dependent, unlike the coupled energy, due to constructive and destructive interference of the various frequencies.

Noise peak voltages increase as a fixed number of logic transitions are concentrated over a narrower time interval even though the overall switching power is fixed and the total coupled noise energy remains unchanged. The switching of a capacitively loaded I/O driver represents an example of concentrating switching power in a narrow time interval. This results in much larger coupled peak noise voltages per milliwatt of switching power for an off-chip I/O driver versus a group of smaller drivers not exactly aligned in phase. In as much that power supply voltage, temperature, and frequency change switching power they also change substrate coupled noise energy. This text refers to p type substrates but the information applies to n substrates with the corresponding bias polarity and junction type reversals.

9.1 Characterization of noise coupling concepts

To demonstrate some of these concepts a test chip was fabricated. The test chip was also fabricated on a degenerately doped P+ substrate with a 10 ohm cm P- epitaxial layer. The test chip had programmable banks of latches and off-chip drivers. The drivers and latch banks were selected or deselected by gating the clocks to them. In addition a substrate bias generator circuit composed of a charge pump was selectable on-chip to study the effect of lowering the substrate bias below ground and to see the effect of the noise. Substrate contacts were present only in a ring contact on the outside perimeter of the chip. Figure 9.2 shows the measured peak to peak voltage noise on the substrate as each of the various circuits is turned on and allowed to switch. Each of the circuits has a different amount of coupled peak to peak voltage noise induced on the substrate per milli-watt of switching power. Since we are measuring peak to peak noise voltages the degree to which the switching is confined to a narrow time interval increases the amount of peak to peak noise voltage.

The first circuit turned on is the substrate charge pump which is a free running VCO (voltage controlled oscillator) that pumps current into a capacitor tied to the substrate to pull it below ground. A diode clamp prevents the substrate from dropping more than 0.6 volts below ground. The noise is coupled to substrate by the switching action of the VCO. The frequency of the VCO in increased to increase the switching power. The noise increases linearly until the diode clamp engages at about 25mw of switching power. The switching frequency of the VCO reached a maximum at about 1Mhz. The next circuit turned on was the clock generation and gating circuitry. This accounts for the single data point on Figure 9.2 at about 70 mw of switching power. The clock frequency is 27Mhz. The following 15 data points reflect the noise and power as 15 identical banks of CMOS static latches are gated on one by one. The noise and power increase is linear but on a different slope than the substrate pump. All of the I/O drivers are turned on next. Note that the I/O drivers are on a higher slope yet due to the fact that they are all highly synchronized and all switch in the same direction at the same time.

It is interesting to note that when the substrate noise voltage waveform was ac coupled, rectified and averaged that the slope of the averaged voltage versus switching power produced a single slope line across all of the switching circuits. This is equivalent to saying that although the coupled noise peak voltage per milliwatt of switching power depends on the circuit; The coupled noise energy per milliwatt of switching power is constant across circuit types in a given environment.

Characterization of noise coupling concepts

The latches used to produce the switching power were a worst case stress for noise since the latches were all synchronized to switch at the same time in the same direction within a 2ns time window after the clock edge. When packaged in the same packaging environment shown in Figure 2.12 the peak to peak voltage noise was: Logic Vdd=1.28volts, Logic ground=1.28volts, Chip Substrate=0.47 volts, Analog Vdd=0.45volts, and Analog ground=0.36volts. Figure 9.1 on page 186 shows this pictorially. When the same switching power was produced by a product design that was composed of a mix of switching circuit functions with switching distributed throughout the period of the 27Mhz clock, the peak to peak noise voltage dropped by three times.

The banks of latches shown in Figure 9.2 were spaced at varying distances away from a number of analog circuits including amplifiers, op-amps, voltage buffers, voltage references etc. Since each bank of latches could be individually selected it was possible to study the effect of how distance from the noise source affected the noise actually coupled to the analog circuit. The banks of logic latches had a very small area of interconnect capacitance since clocks were the only globally distributed nets and they were gated at each latch bank. The output coupled noise to the substrate resulted mainly from the capacitances of the junctions of the switching devices. Since the primary noise coupling mechanism was switching and resonance noise coupled through the substrate and the substrate had a sheet resistance of about 0.1 ohm per square there was basically no difference in turning on the latch bank adjacent to the analog circuit or the one farthest away. This is an idiosyncrasy of degenerately doped P+ substrates.

Various substrate contacts on the test IC showed also that the noise voltage waveform on the substrate did not depend on which single bank of latches were switching relative to the substrate contact. All latch banks were spaced a minimum of 10 microns from any circuit and a maximum of 5 mm away. Reference [9.3] reports that there is some sensitivity when the spacing becomes closer than four times the epitaxial layer thickness which in this case was 6.5 microns.

Controlling Substrate Coupling in Heavily-Doped Bulk Processes

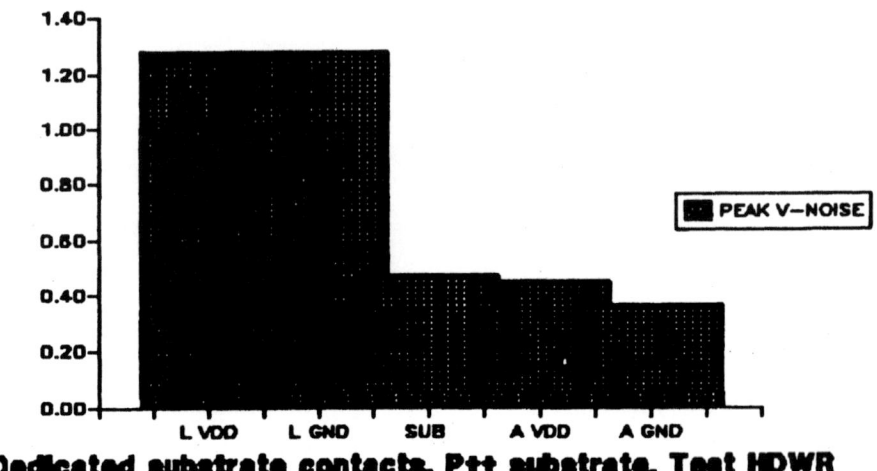

FIGURE 9.1 Relative peak voltage noise.

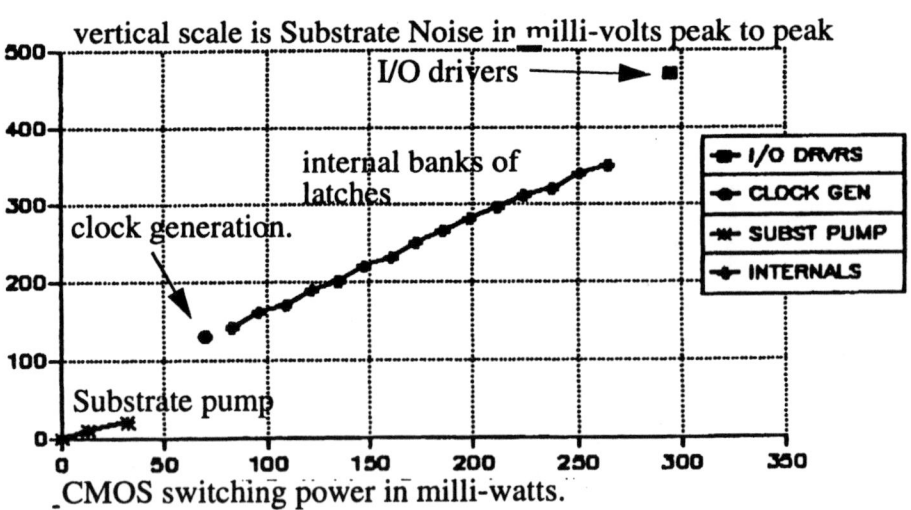

FIGURE 9.2 Substrate Voltage Noise vs CMOS Switching Power.

Characterization of noise coupling concepts

It is interesting to note that not all of the analog circuits tested had a linear relation between the noise resulting on their outputs and the noise on the substrate. This is because some of the circuits were integrating in nature and integrated out the periodic coupled noise and were affected by the non-periodic noise. Other circuits were band limited in frequency and were only sensitive to certain frequency ranges of coupled noise. Still other circuits were fully differential and rejected common mode substrate noise well within their operating bandwidth and outside this bandwidth the rejection decreased. This is another reason why it is important to simulate the circuit in the actual chip/package model environment with the approximated switching disturbances in the proper frequency ranges.

Figure 9.3 shows how the switching frequency affects the switching power and the coupled noise voltage. The bottom trace is merely a verification that the measured circuit power dissipation varies linearly with switching frequency for this design. The top trace shows the peak to peak noise coupled to the chip P+ substrate as the frequency is varied. The switching power was kept constant at each frequency. The value varies with frequency because the phasing of the interference of the clock and resonance frequencies changes with the excitation clock frequency.

When the inductance from chip substrate to card ground was cut in half by the connection of more package pins the voltage peak noise on the substrate dropped to 53% of its previous value. The noise power or energy on the substrate dropped by almost exactly 50%. The difference is because the reduced inductance also changed the resonant frequency and thus the phasing of the voltage interference waveforms. The noise energy is not affected by the phasing of the interference. When the inductance was dropped another 33% the noise peak voltage dropped 26% while the noise energy dropped almost exactly 33%. When the inductance of both switching power rails was cut by a certain percent the coupled noise energy to substrate was cut by almost the same percentage. Likewise when the inductance of both analog power rails was cut by a certain percent the coupled noise energy to the analog rails from the substrate was cut by almost the same percentage.

Controlling Substrate Coupling in Heavily-Doped Bulk Processes

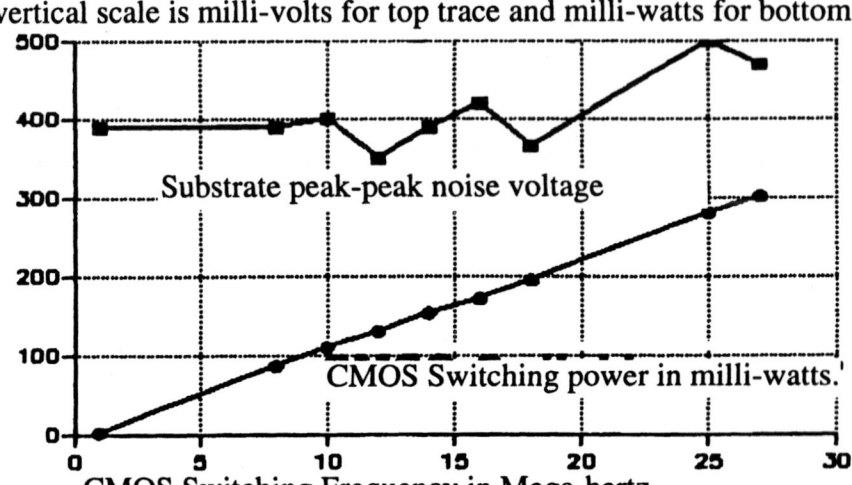

FIGURE 9.3 Effect of Switching frequency on noise and power.

9.2 P+ Bulk Wafer Characterization

Highly doped P+ bulk wafers with grown epitaxial layers are the easiest to analyze because the substrate can usually be considered as a single node, but they usually offer the worst analog (quiet or non-switching circuits) to digital (CMOS or switching circuits) coupling environment. An exception to this exists if the P+ bulk is degenerately doped (~.006 ohm-cm) and can be used as a pseudo ground plane under the epitaxial layer. This usually involves a low inductance and resistance connection to the entire chip backside via a titanium-copper-gold metallization, for example, on p silicon. The wafer must be backside ground to thin the bulk material to be LESS than one skin depth for the entire problem frequency range to make the backside contact an effective ground plane. At one skin depth from the surface of a conductor the current density has dropped to 1/e of its value on the surface or 37%. Making a conductor

P+ Bulk Wafer Characterization

thicker than about two skin depths will have a small effect on the overall resistance. If the backside contact were placed two times the skin depth from the silicon surface at frequency f0 then that backside contact would conduct no signal for frequencies above f0. At 100 Mhz a 0.006 ohm cm, 8 inch, un-thinned wafer will be about two skin depths thick (775 microns). This means that it will conduct no signal current above 100Mhz. It cannot be an effective ground plane above the frequency where the chip or wafer thickness becomes one skin depth or 25Mhz for this example. The wafer can be thinned to about 200 microns thick before the wafer starts to become unmanageably fragile. The frequencies, where the chip thickness of 200 microns is one and two skin depths respectively, are 380Mhz and 1.5Ghz. This is still not an effective backside contact for very high frequencies. Consider the fact that a good square wave carries the first several odd frequency harmonics of the fundamental. A 35Mhz square wave would carry frequency components up to 380Mhz. Because the wafer is being thinned the resistivity of the substrate can be allowed to be somewhat higher. A 0.1 ohm-cm thinned, 200 micron thick substrate would have 0.5 ohms per square lateral resistance, 0.066 ohm vertical resistance (for a 5.5mm square chip), and have its thickness not reach one and two skin depths until 25Ghz and 100Ghz. The equation for skin depth (9.1) is shown below and pictorially in Figure 9.4. The equation for horizontal resistance is (9.3) and vertical resistance is (9.5).

$$T_{skin} = \sqrt{\frac{\rho}{\pi \mu f}} \quad \text{Skin depth} \tag{9.1}$$

$$\mu = permeability$$

for nonmagnetic materials:

$$\mu = \mu_0 = 4\pi \times 10^{-9} \left(\frac{H}{cm}\right)$$

$$\rho = resistivity\,(\Omega - cm) \tag{9.2}$$

making a conductor thicker than about $2T_{skin}$ won't reduce its resistance.

For reference at room temperature aluminum has a 2.8 microns skin depth at 1Ghz.

$$R = \rho \frac{L}{Wt} \quad \text{Horizontal or lateral resistance} \tag{9.3}$$

where: R=resistance in ohms, ρ =resistivity in ohm-cm,

L=length, W=width, t=thickness of wafer.

Controlling Substrate Coupling in Heavily-Doped Bulk Processes

Controlling Substrate Coupling in Heavily-Doped Bulk Processes

- Example of 200 micron wafer that is 0.1 ohm-cm

$$R = \frac{0.1}{0.2} x \frac{L}{W} = 0.5 \frac{(ohm)}{(square)} \qquad (9.4)$$

$$R = \rho \frac{t}{WL} \quad \text{Vertical resistance} \qquad (9.5)$$

- Example of 200 micon thick wafer that is 0.1 ohm-cm and 5.5 mm on a side:

$$R = 0.1 \times \frac{0.2}{0.55 \times 0.55} = 0.066 ohms \qquad (9.6)$$

Figure 9.6 shows the lateral resistance for a uniformly doped bulk silicon if the silicon were contacted on the sides. This assumption is not a bad one for highly doped substrates with very low resistivity and frequencies below the skin depth. As the substrate resistivity increases, such as with 10 ohm-cm bulk P- substrates, the affect of surface point contacts causes the resistance to increase and the current to be concentrated near the silicon surface.

The low inductance connection from a chip backside contact to the card can be achieved with chip-on-board packaging which conductive epoxies the chip backside directly to the card ground. An alternative is conductive epoxy connection of the chip backside to a leadframe die attache paddle that is connected directly to a number of package lug pins. Note that the standard silver epoxy used to attach most chips to package carriers will react away a few tens of angstroms of native oxide on backside ground wafers to produce an inadvertent low ohmic contact on 10 ohm-cm bulk N- substrates. P wafers usually produce a rectifying contact unless they are first metallized. Figure 9.7 shows the effect on substrate noise of having a backside contact. The picture on the left is from simulation using the model of Figure 8.12 for the top trace and the model of Figure 9.28 (backside contact) for the bottom trace. Both traces use the same unbalanced excitation as in the bottom trace of Figure 8.21. The picture on the right shows the measured hardware result. The backside contact parts were fabricated by depositing Ti-Cu-Au on the backside of a thinned chip. The chip was placed in a special PLCC package where the chip die paddle was an aluminum slug that was exposed from the plastic on the bottom side. The chip was attached to the top side of the die paddle with silver filled epoxy. The aluminum die paddle bottom side was plated with Ti-Cu-Au to accept solder and soldered with copper foil to the card ground plane. Resistivity from the P+ silicon bulk to card-ground was measured at 35 milli-ohms. The inductance of the connection was only 65 pico-henry. The substrate noise was reduced roughly two orders of magnitude by this backside contact with a

P+ Bulk Wafer Characterization

100ns clock period. This low clock frequency reduced any skin effects. Figure 9.12 and Figure 9.13 show two possible schemes for chip backside connection to the chip substrate. Figure 9.12 shows the case where the only connection to the chip substrate is the chip backside connection. Figure 9.13 additionally places contacts on all chip ground busses. This allows the backside contact inductance to parallel the top-side ground connection inductance for an overall reduction in ground bus inductance. If the P+ substrate has low enough vertical resistance, the top side ground power package bonds can be eliminated. In this situation the backside contact ties the substrate to ground and provides the path for dc and ac ground current as well. The elimination of top-side ground connections can reduce the size of the overall package which reduces the inductance further. Alternatively, the eliminated top-side bonds can be used to lower the inductance of the Vdd rails as well. Reference [9.9] addresses backside contacts further.

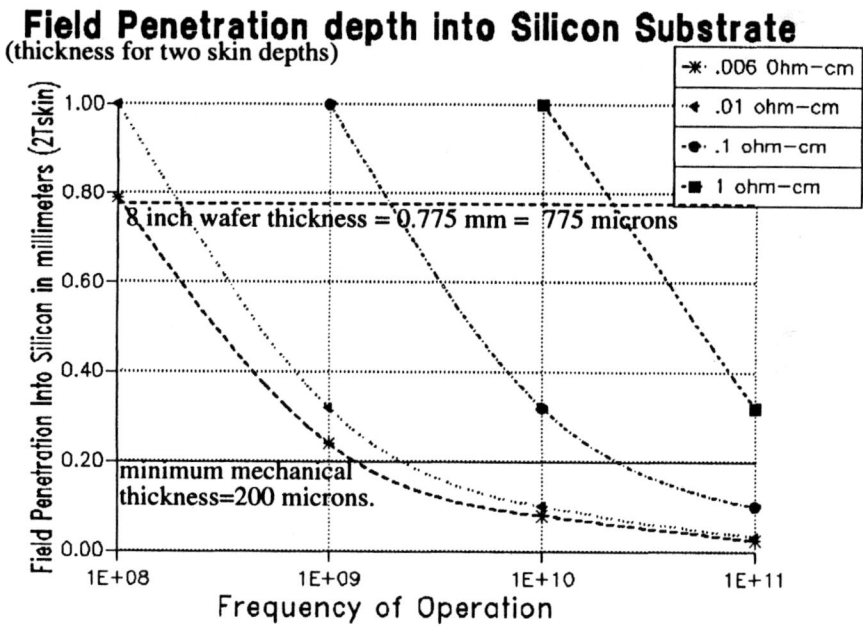

FIGURE 9.4 Necessity to thin wafer for effective high frequency backside connect.

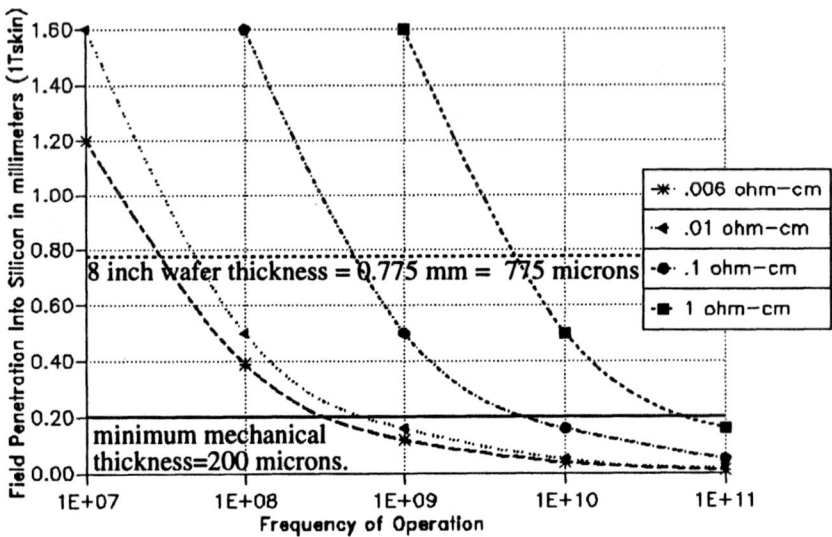

FIGURE 9.5 Single skin depth bounds where backside contact can be a ground plane.

P+ Bulk Wafer Characterization

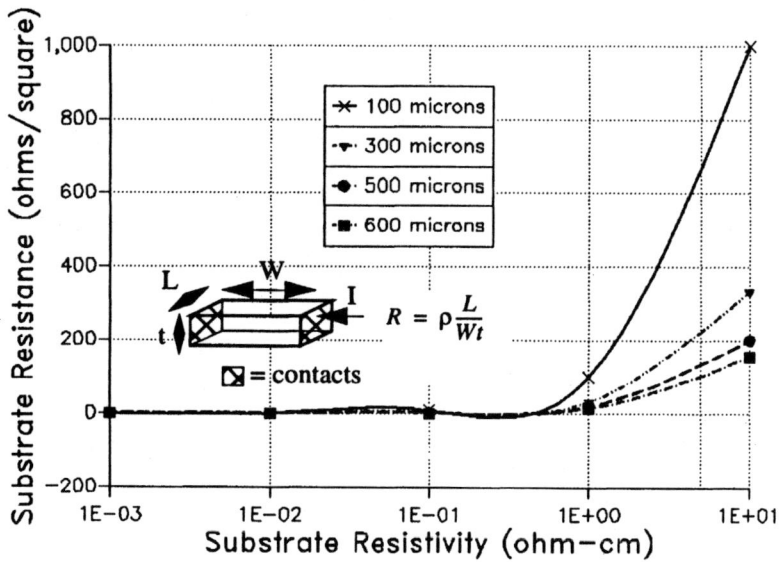

FIGURE 9.6 Lateral resistance per square assuming chip edge contacts.

Controlling Substrate Coupling in Heavily-Doped Bulk Processes

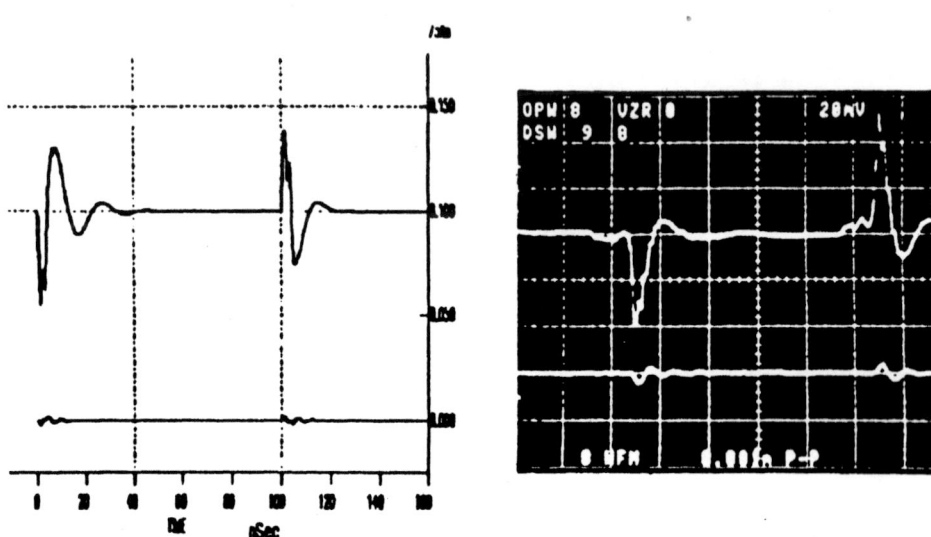

FIGURE 9.7 Substrate waveform without (top) and with(botttom) backside contact.

9.3 Effect of Substrate contact placement on coupled noise

If a backside contact is not used the substrate may be tied via chip top surface bond wires to ground potential. Figure 9.8 shows a situation where the substrate contacts are only placed on the switching ground bus. In most situations this is not the optimum connection scheme since it directly couples the switching power supply bounce into the substrate. Figure 9.9 shows a situation where substrate contacts are placed on both switching and quiet chip ground buses. This is also not optimum since the switching (digital) and non-switching (analog) ground buses are now directly resistively connected through the substrate. This connection resistance can easily be less than one ohm on large chips and a few ohms on small chips. The lower the resistivity of the P+ substrate the better the short between switching and non-switching ground buses. Because of the spreading resistance due to the p channel stop in the p- epi, a

Effect of Substrate contact placement on coupled noise

single few millimeter long and few micron wide contact can make a connection of only a few ohms to the underlying P+ bulk. This enables a few Kelvin contacts not connected to any chip ground to be bonded out to eliminate the need for other substrate ties on chip ground. Kelvin, in this situation, means non-power carrying. Figure 9.10 shows this scheme pictorially. This prevents the direct resistive connection of chip power rails to chip substrate with the remaining substrate coupling mechanism being capacitive. There is no latch-up risk with only a few large substrate ties since the P+ bulk forms a low resistance connection to any place under the p- epi. If extra bond pads are not available to tie the substrate the next best alternative is to tie the substrate to a quiet ground bus. This has the advantage of allowing close source to bulk substrate ties for n-channel devices in the P- epitaxial layer. There is inherent capacitance from the nmos device channel to the bulk substrate as well as body effect modulation of the threshold voltage. Any difference in potential between the ground the n-channel device source is tied to and the ground tied to bulk substrate, increases coupling through the channel capacitance and body effect Vt modulation. If the non-switching or analog ground bus does have some small amount of switching it may prove better to use a separate Kelvin ground for substrate ties. The tying of substrate to switching ground alone is the worst alternative. Figure 9.11 shows a situation where the substrate contacts are only present on the quiet power bus. This is usually close to optimum since only capacitive coupling between switching and non-switching ground buses occurs. Figure 9.8 thru Figure 9.13 show pictorially from worst to best the substrate tie methods for the design of Figure 8.12. Figure 9.14 shows a simulation with the substrate waveforms for the first four approaches with a balanced excitation of the substrate such that only supply resonance noise is present. Putting substrate contacts on the analog power bus only, was better than putting them on a top side Kelvin (non-power carrying) contact because this allowed the extra pins to be assigned to the analog ground bus. Whether this is the order of other designs depends highly on what your model looks like. For example, a recent design included a 16 megabit DRAM on a chip with some analog functions. The DRAM had approximately 37nano-farad of capacitance to substrate from the switching ground bus. For most frequencies of interest the capacitive impedance was so low that the presence of resistive substrate contacts on the switching bus did not make much difference. In fact, placing the substrate contacts on the DRAM ground bus effectively put the nwell to substrate capacitance in the design directly across the power rails for added decoupling. This is an extreme situation but it is best to simulate your design to predict the optimum configuration. It should be noted that the assumption of a single node for the substrate, in the model of Figure 8.12, becomes more justified when substrate contacts are present on the ground power buses. The network of substrate contacts on a widely distributed metal ground bus effectively shorts the substrate epitaxial region under the bus.

9.4 Effect of Package Inductance on Substrate noise

Figure 9.15 shows the RLC resonant substrate noise with substrate ties on all grounds and Figure 9.16 shows the chip RLC (resonant) substrate noise with substrate ties on only the analog or non-switching ground. The figures also show the effect of various package types on this comparison. The waveform for each package type has a DC component added so that the waveforms were separated on the figure. Each waveform on the figures also references one of Figure 8.12 and Figure 9.18 thru Figure 10.1 to indicate the package model used to simulate the corresponding noise. The top trace on Figure 9.15 is for the PLCC package of Figure 8.12. The next trace corresponds to a metallized ceramic Pin Grid Array (PGA) package which has a reduced pin resistance and inductance as shown in Figure 9.18. Note that the low inductance pins close to the chip were used for the power supply pins and not for the substrate tie pin which colors the result somewhat. A typical PGA package is shown in Figure 9.20 and a cross section of the package is shown in Figure 9.21. The third trace from the top on Figure 9.15 corresponds to a metallized ceramic PGA package that has a ground plane in the package. A cross section of the package is shown in Figure 9.22. The ground plane is not electrically tied to any chip pins but rather to a pin tied to card ground to see the effect of inductance reduction. Figure 9.23 shows a blow up the substrate wiring area around the chip. Note that the chip bumps can be placed over active circuitry unlike wirebond pads. The closely spaced pins to the chip were used for power rails. In this case, again the substrate tie pin was not a closely spaced pin. Figure 9.19 shows the electrical model that results. The fourth trace from the top on Figure 9.15 corresponds to a TAB (Tape Automated Bonding) package as shown in Figure 9.25. This example shows a TAB tape with leads in excess of 200 while the comparison was done with a 68 lead TAB tape package. The TAB tape had dual ground planes so that the inductance of the switching and non-switching ground lines could be dramatically reduced. Figure 9.24 shows the electrical model that results. The fifth trace from the top on Figure 9.15 corresponds to an IC that was bumped with solder and flip-chip connected to a circuit card. Figure 9.26 on page 212 shows the electrical model and Figure 9.27 shows a cross section of a flip-chip on board package. The sixth trace from the top on Figure 9.15 corresponds to a PLCC packaged chip with a very low inductance connection from the backside of the chip to the circuit card. Figure 9.28 shows the electrical model. The seventh trace from the top on Figure 9.15 corresponds to a chip packaged in a standard PLCC package but with a P- substrate that allows approximately 50 ohms of resistance to be obtained between the switching and non-switching ground buses. The electrical model is shown in Figure 10.1 and the topic of P-substrates will be taken up in a later section. Note that the bottom trace on both Figure 9.15 and Figure 9.16 uses substrate contacts on both the switching and non-switching ground buses for latch-up reasons.

Effect of Package Inductance on Substrate noise

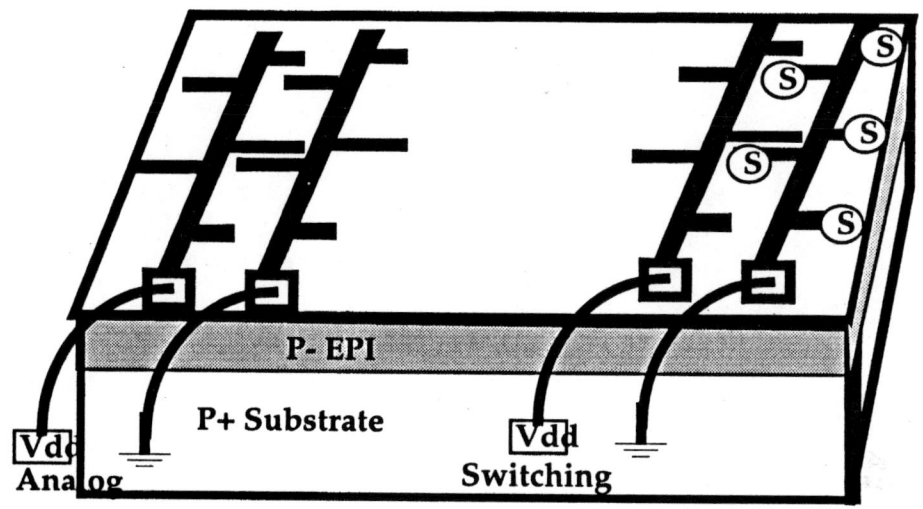

FIGURE 9.8 Substrate Contacts on Switching ground. (usually not recommended!)

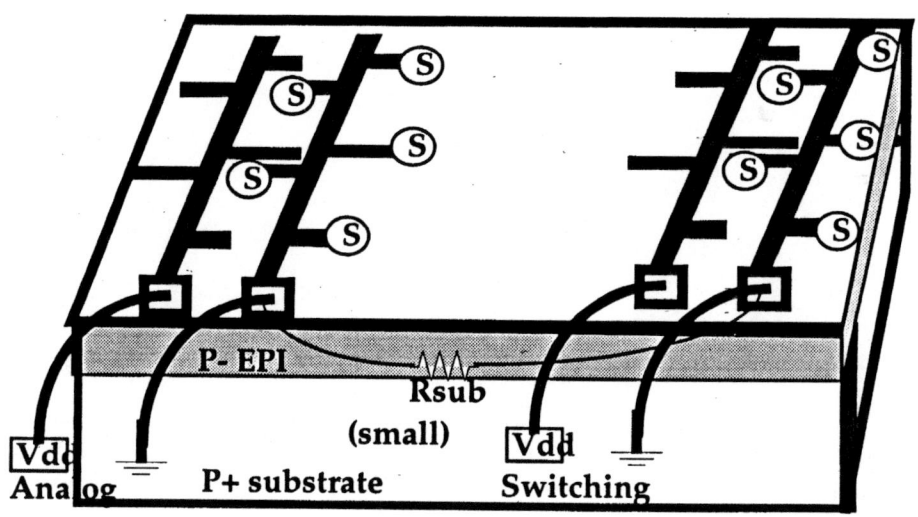

FIGURE 9.9 Substrate contacts on all ground busses; Forms resistive short!

Controlling Substrate Coupling in Heavily-Doped Bulk Processes

Controlling Substrate Coupling in Heavily-Doped Bulk Processes

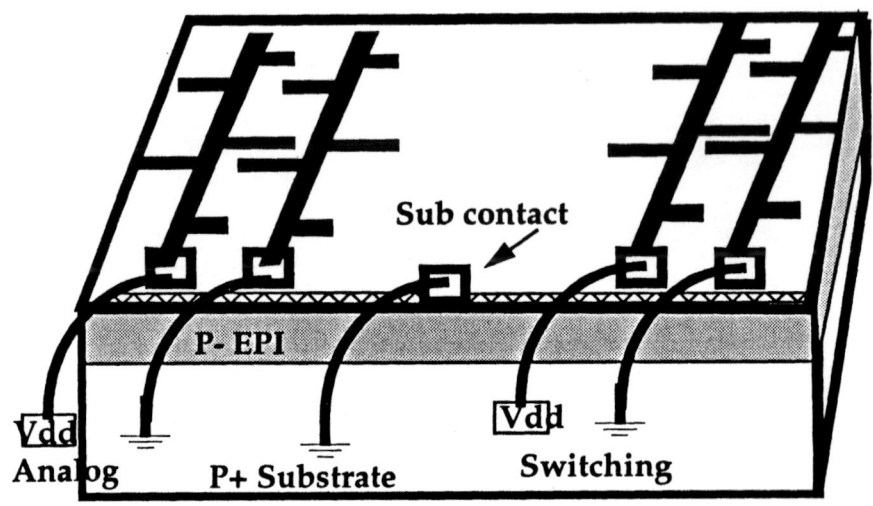

FIGURE 9.10 Substrate Contact to Kelvin Top-Side Ground. (good!)

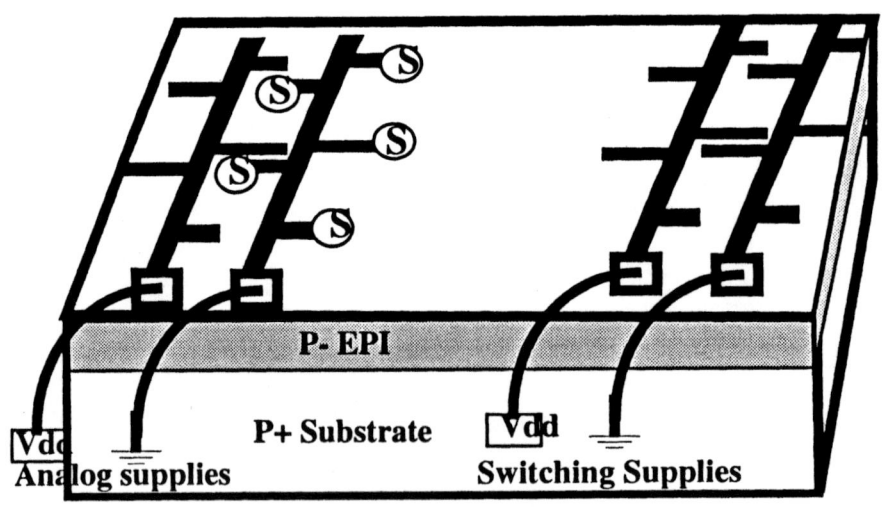

FIGURE 9.11 Substrate Contacts on Non-Switching Ground. (good!)

Effect of Package Inductance on Substrate noise

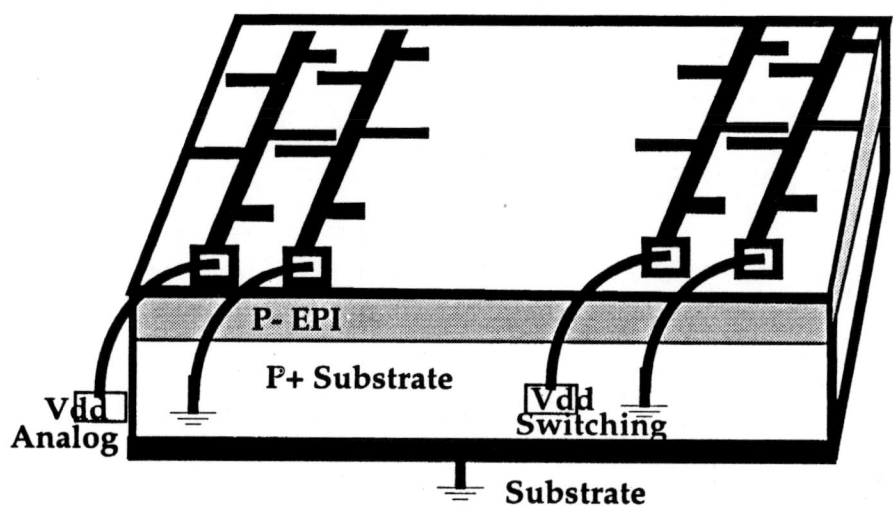

FIGURE 9.12 Substrate Contact on Kelvin (non power carrying) backside contact.

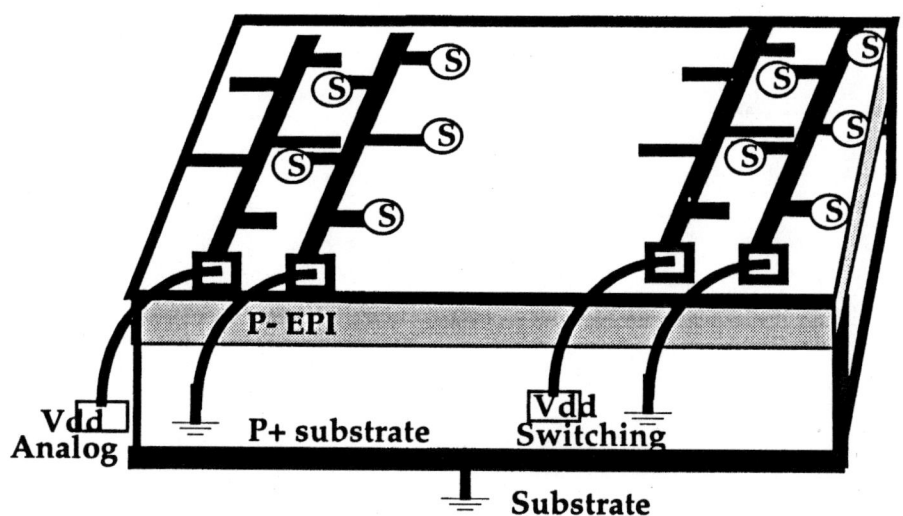

FIGURE 9.13 Substrate Contact on power carrying backside contact.

Controlling Substrate Coupling in Heavily-Doped Bulk Processes

Controlling Substrate Coupling in Heavily-Doped Bulk Processes

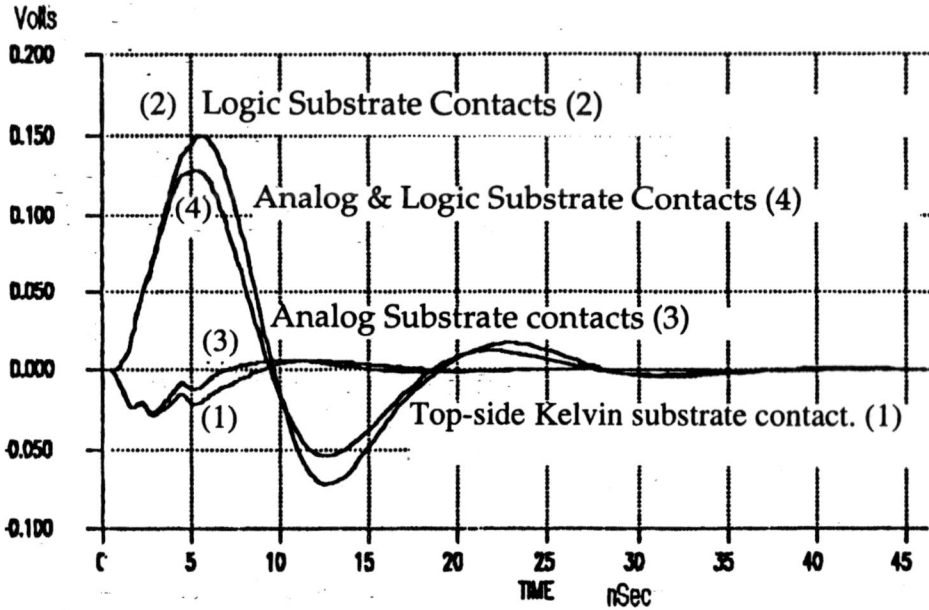

FIGURE 9.14 Where to put substrate contacts for model of Figure 8.12.

Table 9.1 and Table 9.2 give simulations corroborated with hardware showing chip substrate, analog and digital supply noise for the IC of reference [9.2]. The various packaging schemes show the effect of the inductance variation. The packages are shown top to bottom from highest to lowest inductance. Table 9.2 also gives the noise effect from "unbalanced" switching and I/O driver switching which couples noise directly to the chip substrate as well as exciting the chip/package resonance.

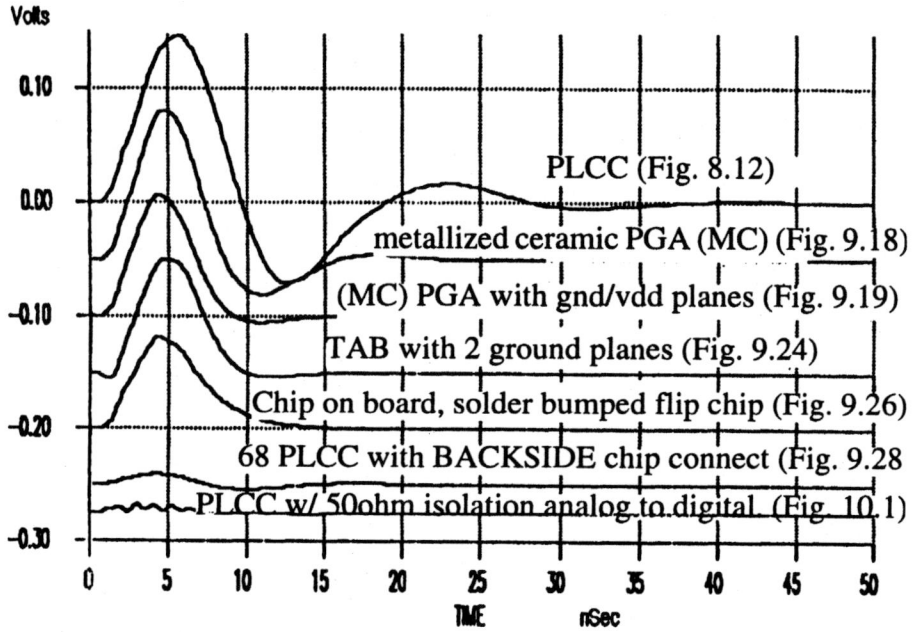

FIGURE 9.15 Substrate RLC noise with substrate ties on all grounds.

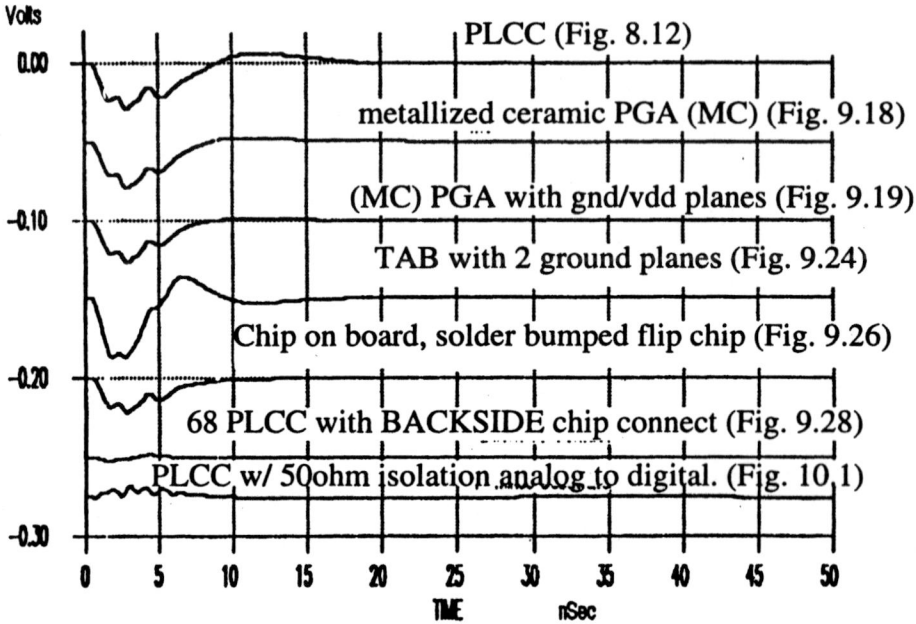

FIGURE 9.16 Substrate RLC noise with only analog substrate contacts.

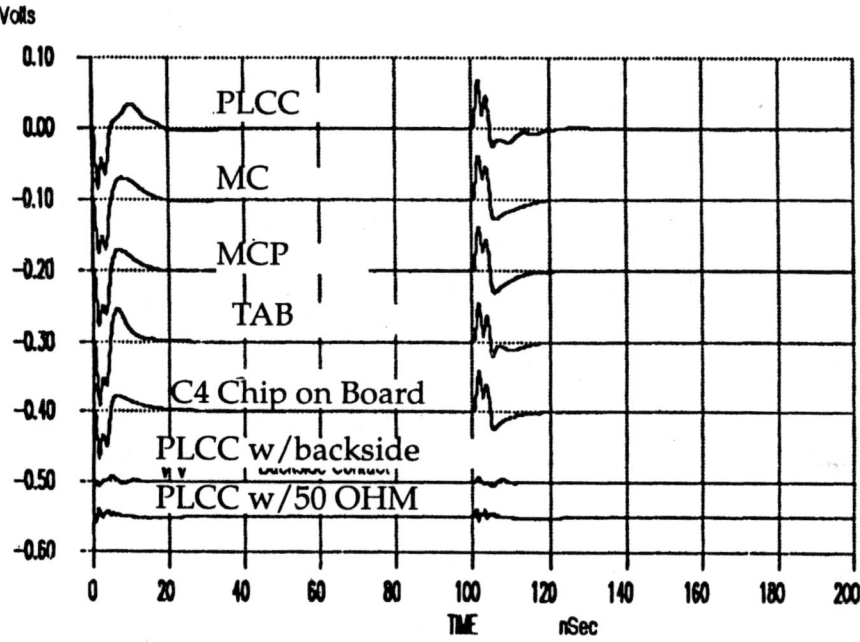

FIGURE 9.17 Substrate output-coupled noise with analog sub contacts.

9.5 Noise Coupling Control Techniques

Experiments with P+ bulk/p- epi wafers show that separation of noise transmitter from noise receiver makes almost no difference in coupled noise due to the low resistance bulk substrate. The substrate's low resistance makes most current flow through the epi region vertically and renders nwell guard rings basically ineffective for shielding a device from other than its nearest neighbor. P+ guard rings when spaced close to the epi thickness away and tied to a non contaminated ground can reduce substrate coupled noise to an nfet as much as 30%.[9.3] The prime coupling mechanisms are body effect and capacitive coupling. To first order the coupling from substrate to a circuit node is proportional to the capacitance to substrate from that node and the impedance of the coupled node. $V_{coupled} \propto V_{substrate} \times C_{(node) - (substrate)} \times Z_{node}$
A guard ring that reduces capacitance to substrate from a device and stays non-con-

taminated due to coupling from the substrate reduces coupling to the device proportional to the capacitance reduction. As a result substrate noise coupled to pfets in an nwell tied to low inductance power will be lower than for a nfet in the epi. Because of the large nwell area with the substrate and the large capacitance between the power rails it will be difficult to get a "quiet" tie point for a large number of nwells. Often a separate bond wire may be required to obtain a quiet tie point for each group of circuitry. Likewise, a polysilicon resistor will have less noise coupled from substrate than a diffused resistor in the epitaxial layer. The most effective noise control techniques with this substrate type are minimizing power and substrate inductance, putting substrate ties on one quiet bus, tuning power resonant frequencies away from frequencies of interest and optimally, physically partitioning the power and associated package pins among quiet and switching functions.

TABLE 9.1 CMOS logic noise with substrate ties on all grounds/analog ground.

Package, all are P+sub except last	substrate noise (mv)	logic-Vdd noise (mv)	logic-gnd noise (mv)	analog-Vdd noise (mv)	analog-gnd noise (mv)
PLCC-68 pin	218 / 34.8	267 / 318	244 / 297	103 /42.9	113 /44.8
metallized ceramic	160 / 31.4	189 / 227	180 / 215	36.6/19.3	40.4/20.4
multi-layer ceramic	112 / 27.9	141 / 157	127 / 153	22.4/13.8	23.5/14.7
TAB tape	104 / 50.9	170 / 185	118 / 139	24.5/20.1	20.6/16.6
bumped flip chip Cob	81.5/22.2	117 / 120	98.2/ 118	9.9/ 6	9.2/ 5.6
backside contact inside a PLCC	13.1/ 3.9	402 / 337	121 / 287	11.5/ 5.5	12.9/ 5.8
PLCC,P-bulk with 50 ohm analog/ logic separation	8.0/ 8.0	267 / 267	266 / 266	8.4/ 8.4	7.2/ 7.2

TABLE 9.2 CMOS I/O driver noise (Logic ground bus not tied to chip substrate).

package, P+sub except last	substrate noise (mv)	logic-Vdd noise (mv)	logic-gnd noise (mv)	analog-Vdd noise (mv)	analog-gnd noise (mv)
PLCC-68 pin	152	363	360	139	142
metallized ceramic	136	237	237	83.1	90.3
multi-layer ceramic	138	176	178	67	70.6
TAB tape	148	197	164	58.2	52.8
bumped flip chip	124	128	129	28.5	26.7

Noise Coupling Control Techniques

TABLE 9.2 CMOS I/O driver noise (Logic ground bus not tied to chip substrate).

package, P+sub except last	substrate noise (mv)	logic-Vdd noise (mv)	logic-gnd noise (mv)	analog-Vdd noise (mv)	analog-gnd noise (mv)
backside contact inside PLCC	14	420	403	22.1	23.4
PLCC, P-bulk with 50 ohm analog/digital separation	18.3	271	269	18.4	18.4

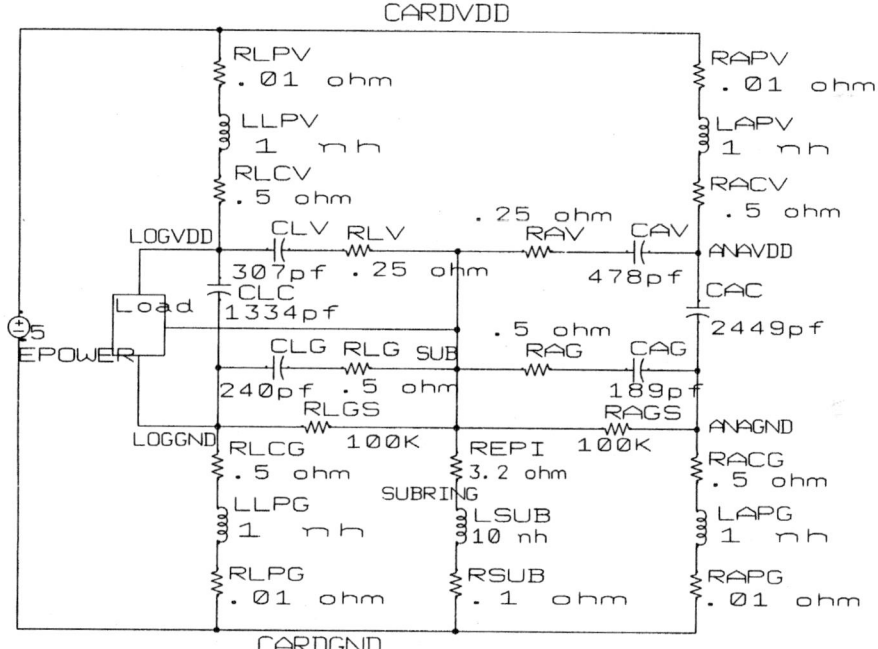

(Packaged in a metallized ceramic pin grid array)

FIGURE 9.18 P+ substrate model for metallized ceramic pin grid array (PGA).

Controlling Substrate Coupling in Heavily-Doped Bulk Processes

Controlling Substrate Coupling in Heavily-Doped Bulk Processes

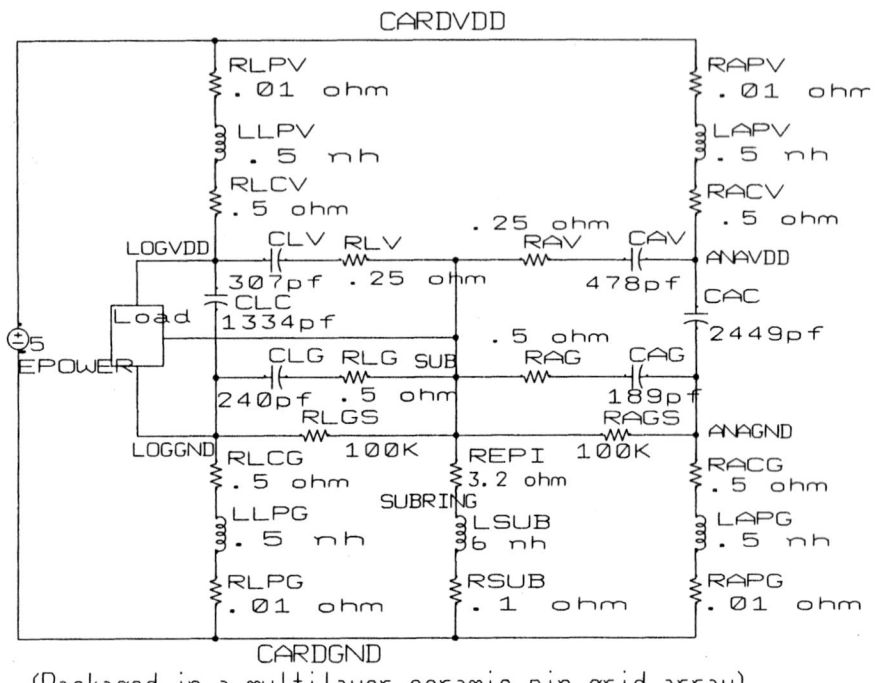

FIGURE 9.19 **P+ substrate model for ceramic PGA with power planes.**

Noise Coupling Control Techniques

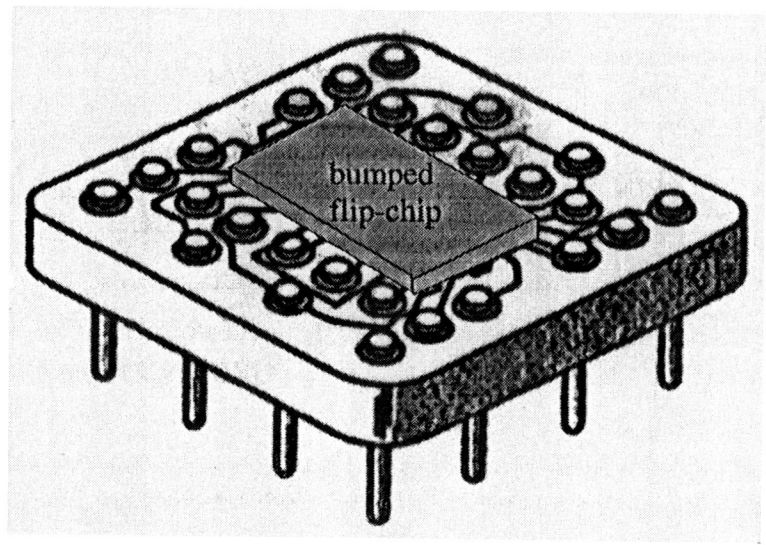

FIGURE 9.20 Drawing of a metallized ceramic (MC) Pin-Grid-Array (PGA).

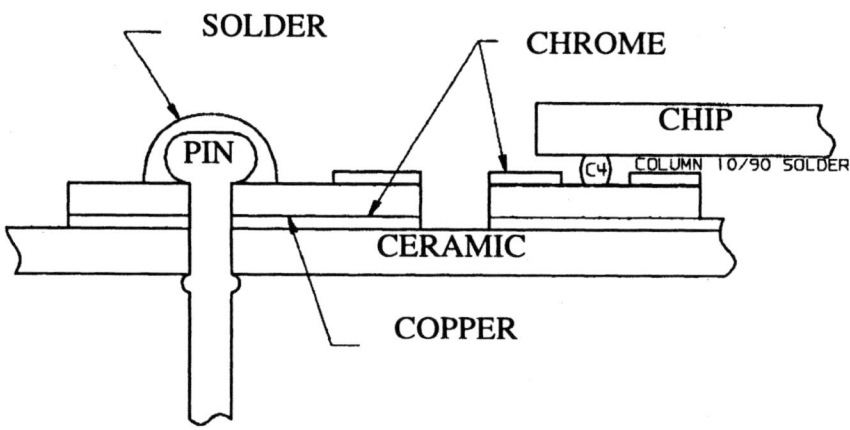

FIGURE 9.21 Drawing cross-section of metallized ceramic (MC) Pin-Grid-Array.

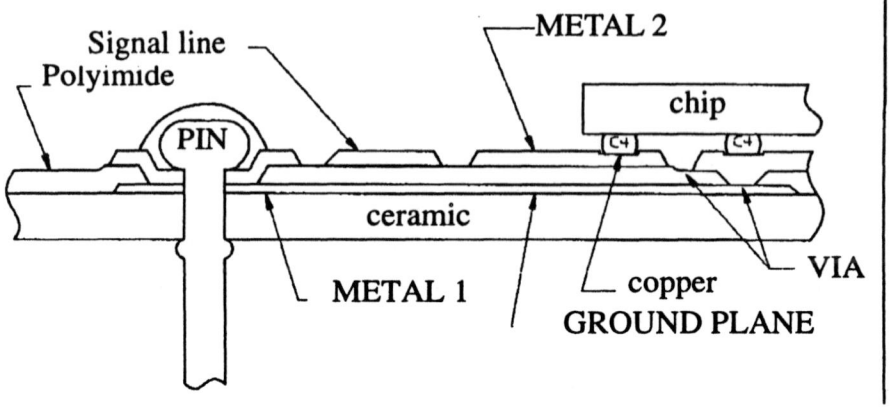

FIGURE 9.22 Drawing cross-section of multi-layer metallized ceramic PGA.

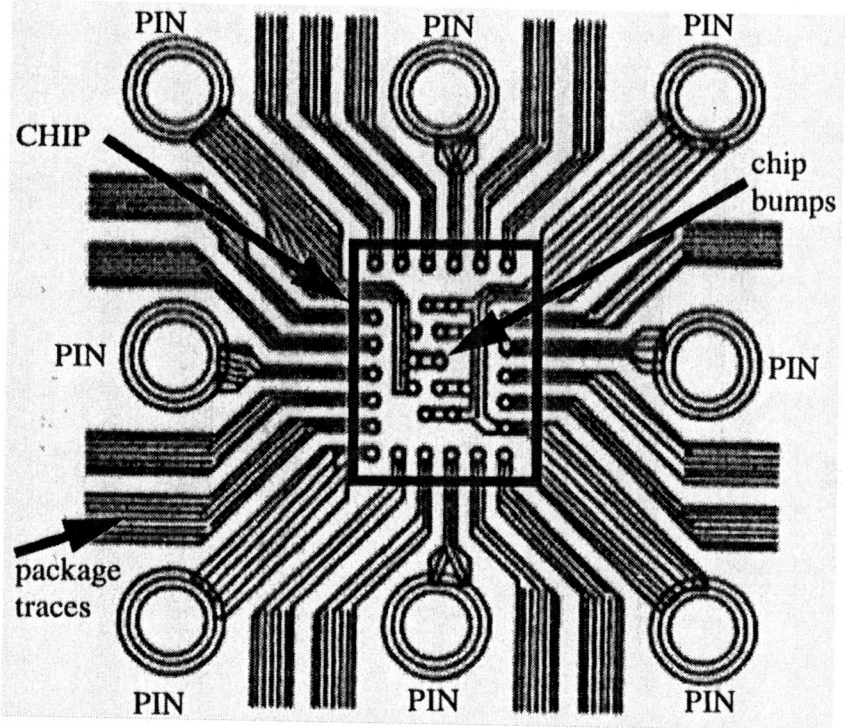

FIGURE 9.23 Blow-up of wiring from under flip-chip for MC/MCP PGA module.

Controlling Substrate Coupling in Heavily-Doped Bulk Processes

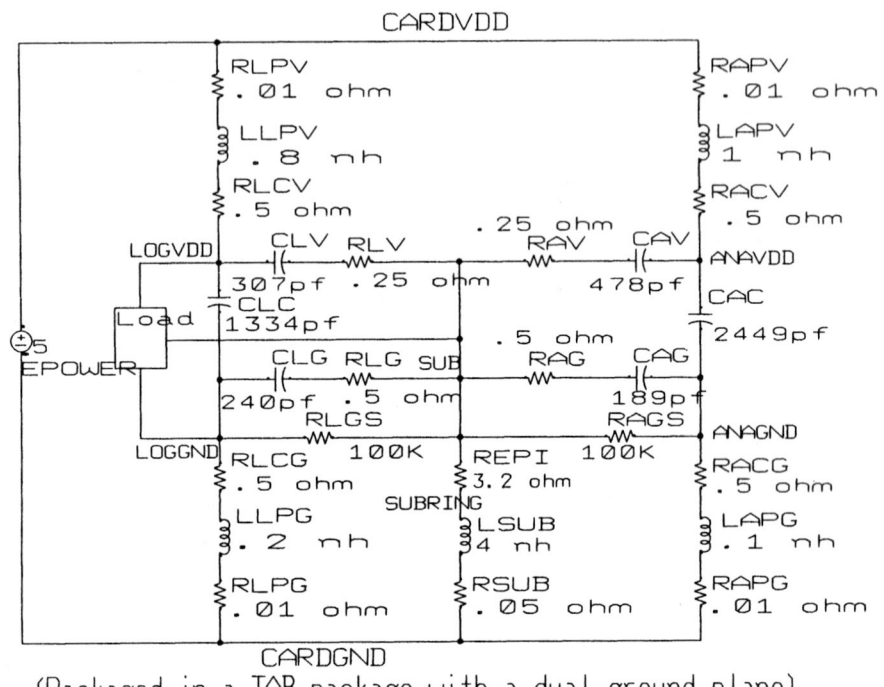

(Packaged in a TAB package with a dual ground plane)

FIGURE 9.24 P+ substrate model for TAB package with dual ground plane.

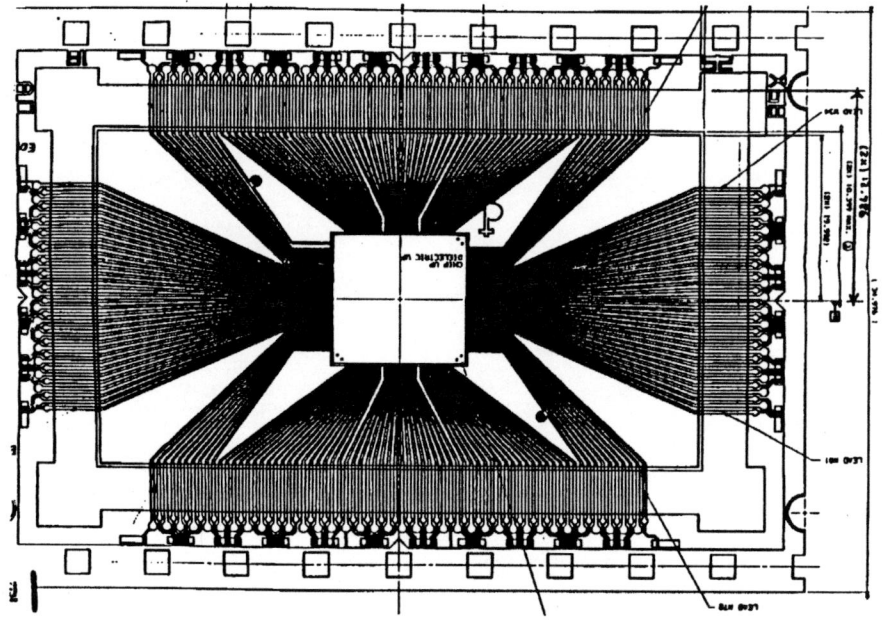

FIGURE 9.25 TAB Package prior to being separated from the tape carrier.

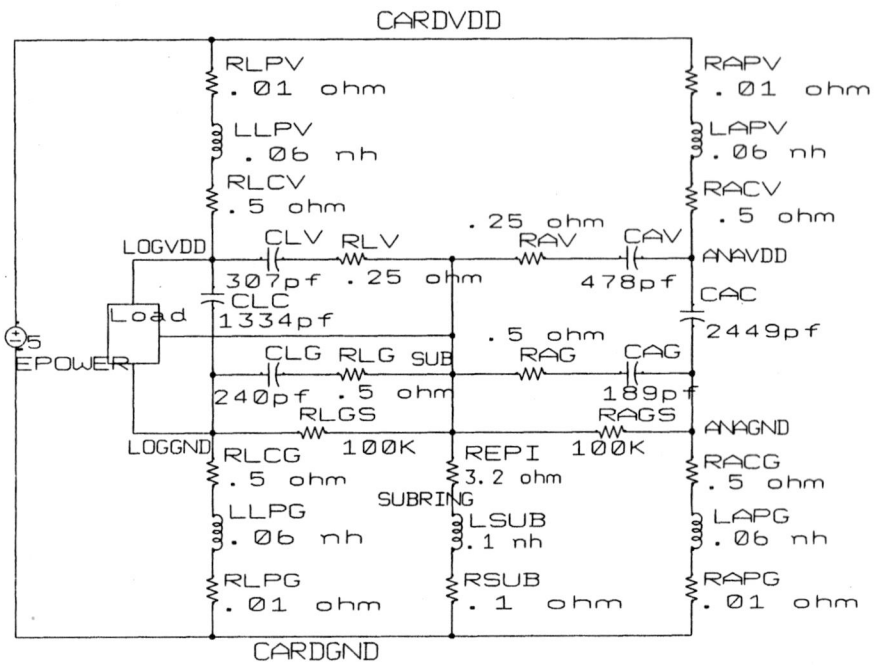

FIGURE 9.26 P+ substrate model for C4 flip chip or chip on board connection.

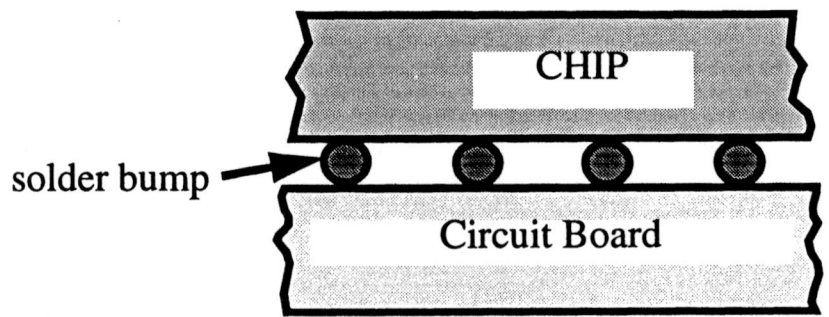

FIGURE 9.27 Cross-section of a flip-chip on board packaging scheme.

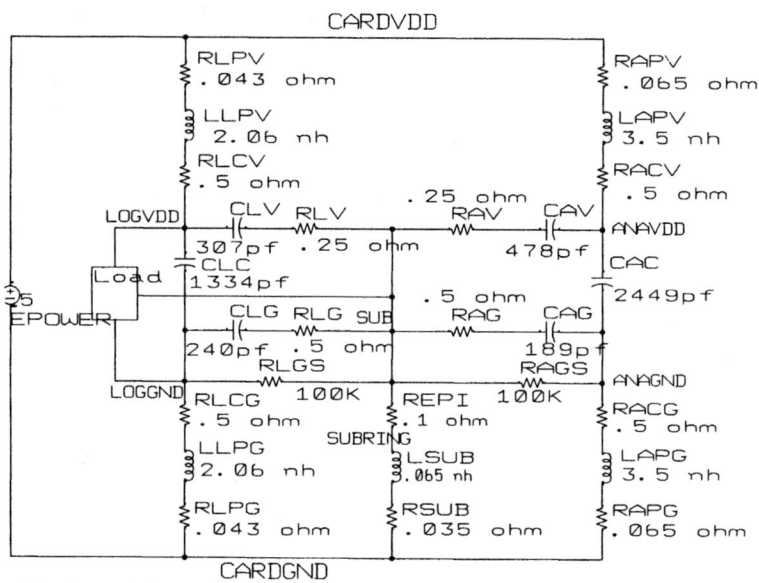

(Packaged in a 68 PLCC with the substrate tied to the heat spreader)
(The heat spreader is tied to card ground through soldered copper foil)

FIGURE 9.28 P+ substrate model for 68PLCC with back side chip and card attache.

Tradition C4 Chip Attach   Low Temperature Flip Chip on FR-4

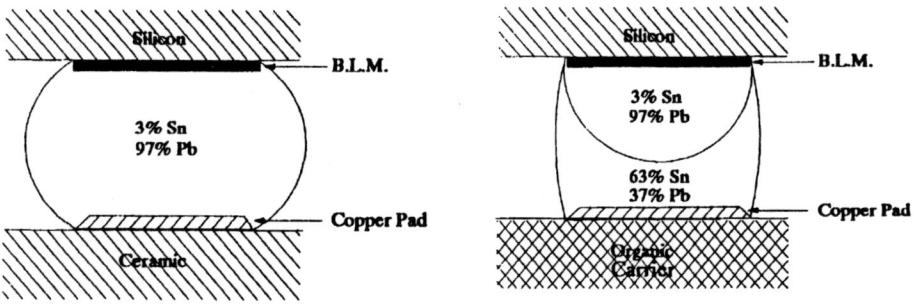

FIGURE 9.29 C4 Ball difference between chip on FR4 verses chip on ceramic.

9.6 Summary

This chapter surveyed techniques and strategies for minimizing coupling through the chip substrate on mixed-signal IC's. Consideration was given to substrate coupling, power rail and i/o driver resonances, near field capacitive and inductive as well as package interaction. Special attention was given to distribution of substrate contacts and how the substrate is referenced in highly doped substrates. The next chapter will deal with lightly doped to insulating substrates.

REFERENCES

[9.1] Timothy Schmerbeck, "Mechanisms and Effects of Noise Coupling in Mixed Signal ICs", EPFL, Switzerland course presentation, June 29-July 10, 1992.

[9.2] T. Schmerbeck, et al., "A 27 Mhz Mixed-Signal Magnetic Recording Channel DSP Using PRML," *Technical Digest of IEEE ISSCC*, pp. 136-137, Feb. 1991.

[9.3] D.K. Su, M.J. Loinaz, S. Masui and B.A. Wooley, "Experimental Results and Modeling Techniques for Substrate Noise in Mixed-Signal Integrated Circuits," *IEEE Journal of Solid State Circuits*, Vol. 28, No.4, pp. 420-430, April 1993.

[9.4] L. D. Smith, *et al.*, "A CMOS-Based Analog Standard Cell Product Family", *IEEE Journal of Solid-State Circuits*, Vol. 24, No. 2, pp. 370-379, April 1989.

[9.5] L. D. Smith, T. Schmerbeck, et al., "A CMOS-Based Analog Standard Cell," *IEEE Journal of Solid-State Circuits*, vol. 24, No. 2, pp. April 1989.

[9.6] H. B. Bakoglu, Circuits, Interconnections, and Packaging for VLSI, Addison-Wesley, Reading, Massachusetts, 1990

[9.7] Charles S. Walker, Capacitance, Inductance and Crosstalk Analysis, Artech House, Boston, 1990.

Summary

[9.8] Notes for the IEEE SSCTC workshop on NOISE in MIXED A/D ICs, held in Williamsburg Va, Sept 6-7, 1990. Tim Schmerbeck and Larry Smith. There were no proceedings,

[9.9] T. Gabara, "Reduced Ground Bounce and Improved Latch-Up Through Substrate Conduction," *IEEE Journal of Solid-State Circuits*, 23, No. 5, pp. 1224-1232, October 1988.

[9.10] Henry W. Ott, "High Speed Digital Design," Electromagnetic Compatibility Seminar, Henry Ott Consultants, January 1992.

[9.11] Dr. Tom Van Doren, "Grounding and Shielding Electronic Systems," NTU Satellite Network, February 1991.

[9.12] R. Philpott, T. Schmerbeck, et al., "A 7Mbyte/sec (65Mhz), Mixed Signal Magnetic Recording Channel DSP Using Partial Response Signalling with Maximum Likelihood Detection," *IEEE Custom Integrated Circuits Conference*, May 1993.

[9.13] N. Verghese, D. Allstot, and S. Masui, "Rapid Simulation of Substrate Coupling Effects in Mixed-mode IC's," In *Proceedings IEEE Custom Integrated Circuits Conference*, pp. 18.3.1-18.3.4, May 1993.

[9.14] T. Schmerbeck, "Design Strategies for Reducing the Effects of Noise Coupling in Analog and Mixed-Mode ICs," in presentation for course on *Practical Aspects of Analog and Mixed-Mode IC Design*, Beaverton Oregon, May 18, 1993.

[9.15] B. R. Stanisic, R. A. Rutenbar, and L. R. Carley, "Power Distribution Synthesis for Analog and Mixed-Signal ASICs in RAIL," In *Proceedings IEEE Custom Integrated Circuits Conference*, pp. 17.4.1 - 17.4.5, May 1993.

[9.16] Internal IBM Technical Report by Larry Smith, IBM, San Jose, CA.

[9.17] Internal IBM Technical Report by Tim Schmerbeck, IBM, Rochester, MN.

[9.18] Kazuo Kato, Hideo Sato, Yasuji Kamata, Kenkichi Yamashita, and Seiichi Ueda, "A 300-Mhz Monolithic Video Current Driver for High-Resolution CRT Applications," *IEEE Journal of Solid State Circuits*, Vol. 24, No.4, pp. 1110-1117, August 1989.

CHAPTER 10

Controlling Substrate Coupling in Bulk P- Wafers

10.1 Bulk P- Wafer Characteristics

Bulk P- wafer CMOS and BiCMOS processes are the most common and inexpensive in use today. Typical resistivity is 10 ohm-cm and ties to substrate are required to be distributed across the chip to prevent latch-up. NMOS and Bipolar processes do not require distributed substrate ties for latch-up protection and thus have more freedom in controlling substrate noise. Logic families found in NMOS and Bipolar processes are generally less noisy than those found in CMOS processes. Some CMOS processes use p+ buried layers to relax the spacing requirements of substrate contacts to prevent latchup, but the treatment is essentially the same as for standard CMOS processes. Backside contacts are not sufficient to prevent against latch-up alone even when the chip is thinned to 200 microns thick.

Table 10.1 shows some scaling relationships and emphasizes that as device dimensions scale down by the factor S that substrate doping will scale up by S. A 4 times dimension scale down (two process generations) of current 10 ohm-cm processes will bring substrate resistivity to around 2 ohm-cm causing bulk P- substrates to look more like P+bulk/P- epitaxial substrates. This will also require more chip area be used to separate switching and non-switching power domains to achieve a given value of separation resistance. Figure 10.6 shows substrate contacts on both switching and non-switching grounds with the two power domains separated by a nwell that breaks the surface field implant and increases the resistance between the two power domains.

Controlling Substrate Coupling in Bulk P- Wafers

Figure 10.1 shows the electrical model for this configuration where it is assumed that the resistance between the two power domains is 50 ohms. Also the nwell separating the two power domains is not floating but tied to the analog or non-switching positive power supply rail. The capacitance of this nwell is modelled as CLS and CAS. It is possible to float the isolation nwell but the floating nwell will float to a low forward bias junction condition and result in a very high junction capacitance shunting the isolation resistance. This model is similar to the model of Figure 8.12 except that the single node substrate is replaced by two single node substrates separated by 50 ohms in parallel with the nwell shunt capacitance. The assumption of a single node substrate in each of the two power domains is very accurate as long as each region has a distributed network of substrate ties shorted together by a metal bus. A large wide nwell is commonly placed around partitioned non-switching power areas of the chip. The nwell breaks the p surface channel stop to provide a higher resistance path between the switching and non-switching power domains on chip. The nwell must also extend outside the chip on its edges into the dicing or scribe area between chips otherwise a lower resistance path between power domains can be inadvertently obtained. This isolation or nwell region encircling the power domain is often referred to as a "moat" in much the same way that moats in medieval times circled castles for isolation and protection. Reference [10.18] used substrate separation to design a 300Mhz video current driver in 1989. Figure 9.15 thru Figure 9.17; Table 9.1 and Table 9.2 show that this noise isolation approach is equal or superior to any packaging or isolation approaches used with P+ substrates. Since P- substrates are also less expensive than P+ substrates with P- epitaxial layers they are preferred in mixed analog and digital (mixed signal) designs. Superior isolation can be obtained, however, with silicon on insulator approaches at somewhat higher cost.

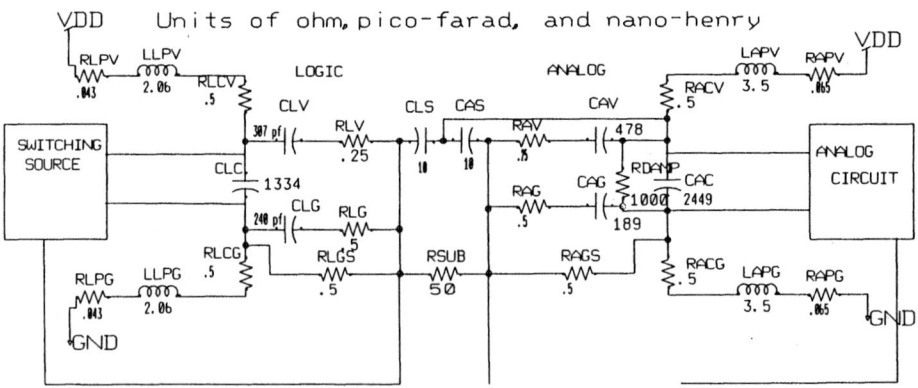

FIGURE 10.1 Model for P- substrate.

Bulk P- Wafer Characteristics

TABLE 10.1 Implications of Scaling on Substrate Doping.

SCALING PARAMETER	
Internal field strengths	1
Velocity saturation, oxide breakdown carrier multiplication effects unchanged	1
Dimensions W, L, tgox, Xj	1/S
Substrate Doping Nsub	S
Voltage Vdd, Vtn, Vtp	1/S
Current per device Ids	1/S
Gate capacitance Cg=eox*WL/tox	1/S
Transistor on resistance Rtr~~ Vdd/Ids	1
Intrinsic gate delay tau=Cg*DELV/Iav=Rtr*Cg	1/S
Power dissipation per gate P=IV	1/S**2
Power delay product per gate P*tau	1/S**3
Area per device A=W*L	1/S**2
Power dissipation density P/A	1

Because of the area penalty to route a separate substrate tie bus and the desire to keep source to body ties very close together to reduce noise in MOS devices, substrate contacts are usually tied to each ground bus on the chip. Figure 10.7 shows an example of using a separate Kelvin ground to tie the surface substrate contacts to. Note that when a single Kelvin substrate bus is used for switching and non-switching power regions of the substrate, that it shorts out the entire substrate and aids in coupling between the two power domains. It essentially creates a single node substrate for the entire chip. To prevent this shorting of the chip substrate, a separate Kelvin bus for tying substrate contacts must be provided in each power domain. Because this requires large amount of chip wiring area, it is not often used. This extensive region of Kelvin substrate ties is also very susceptible to coupling from the power rails and signal wires that run adjacent to it. Spacing the substrate tie bus from other signals further increases its area penalty.

Figure 10.8 shows the use of a single backside contact to provide substrate contacts. The vertical resistance to the backside contact must be low enough to prevent latch-up problems which usually means less than 1 ohm-cm substrates. Also, with a backside contact the current flow will be vertical in the substrate so any surface horizontal separation of switching from non-switching power domains will be "shorted out" by the effect of the back-side contact. This is essentially the same situation as with P+ substrates except the substrate resistance makes the backside contact a poor ground plane and increases coupling.

The last row of Table 9.1 and Table 9.2 show comparisons of power supply and substrate coupled noise when 50 ohms resistance is obtained between the switching and non-switching power domains. Note that when the chip substrate is split, in this way, that a large portion of the return currents flow off-chip since this is a lower impedance path. Often the currents are forced to partially flow within the package by bonding both power domain ground leadframe pins to the chip die attache under the chip reducing the loop area for radiated emissions.

Measurements and simulations of less than 6mm square chips show that, in a 10 ohm-cm bulk silicon substrate, the majority of substrate current flows within about 200 microns from the chip surface. The farther the source and sink of substrate currents are separated the deeper the current will flow. Groups of substrate contacts shorted together through a metal bus tend to shorten the distance between source and sink and cause currents to flow more shallowly. Two power domains separated by 100 microns of distance at their edges will have very shallow current flow in the substrate between them if both domains have a distributed network of substrate contacts shorted together by the metal of their respective ground busses. The two domains may be very large but the current flow will be predominantly in the ground power busses and enter the substrate shallowly only for the 100 micron separation region between them. The effective source to sink separation is only the separation between the two power domains at their edges.

The p channel stop at the silicon surface has doping levels of typically three or four orders of magnitude higher than the bulk substrate so much current flow is concentrated at the surface. This also means that larger separation distances are required to obtain a given value of separation resistance. Because of this, the surface channel stop implant must be blocked between the switching and non-switching power domains. Figure 10.2 shows a typical modern BiCMOS process implant profile. This process also has a P+ buried layer to improve latch-up performance. Note that the surface channel stop implant under the field oxide is doped 3.5 orders of magnitude higher than the bulk substrate.

Bulk P- Wafer Characteristics

Figure 10.3 shows pictorially the various methods of blocking the surface channel stop implant with their corresponding advantages and disadvantages. The methods in order from left to right are; Block all implants to produce native substrate; Use an nwell implant to block the surface channel stop implant; Use an nwell with a subcollector implant to drive the reverse bias layer deeper and lengthen the effective distance between the contacts; and finally allow the surface channel stop implant to remain for comparison purposes. Note that for the 1st case where native substrate remains in the moat isolation layer that the field oxide threshold over this region may be less than the power supply voltage. This will mean that wires may not cross this region on polysilicon or 1st layer metal. Often this can be accommodated if not many wires cross the isolation region and more than one level of metal is used in the design. The moat resistance in ohm-cm when x=100 microns is shown for each structure along with its corresponding -3db roll-off frequency of the isolation resistance.

Figure 10.4 shows the top surface view of the moat region. The simplifying assumption was made, that since the current is concentrated at the surface, the contact could be approximated as on the sides of the P+ contact implant.

Figure 10.5 shows how the resistance of the moat varies with separation distance and method of forming the moat isolation region. The topology of Figure 10.3 and Figure 10.4 is assumed. The moat separation resistance is given in ohm-cm and needs to be divided by the moat width to obtain the separation resistance in ohms.

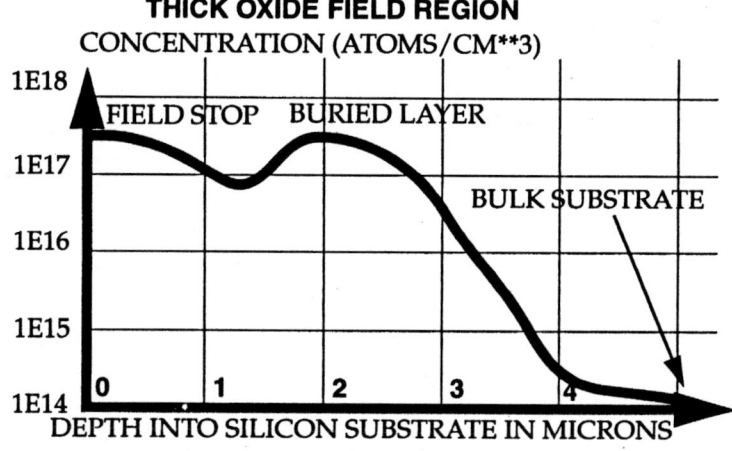

FIGURE 10.2 Typical BiCMOS implant profile.

Controlling Substrate Coupling in Bulk P- Wafers   **221**

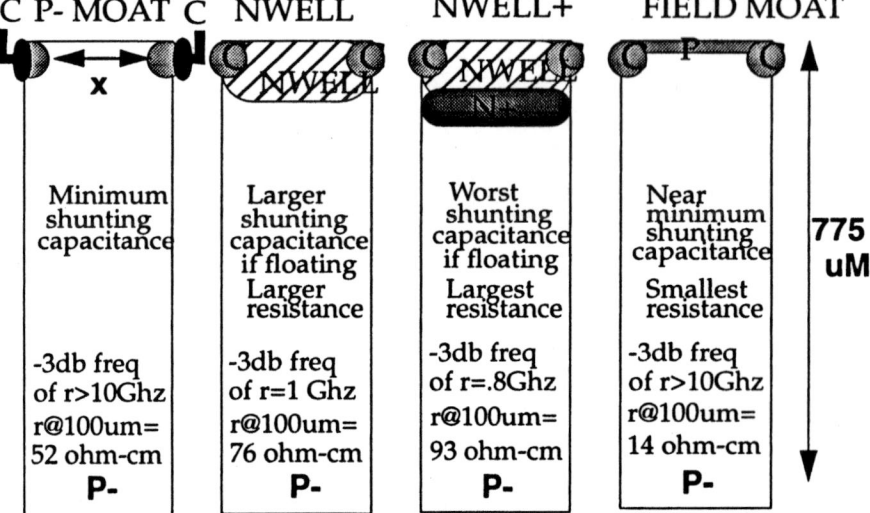

FIGURE 10.3 Performance of different means to separate analog from digital on chip.

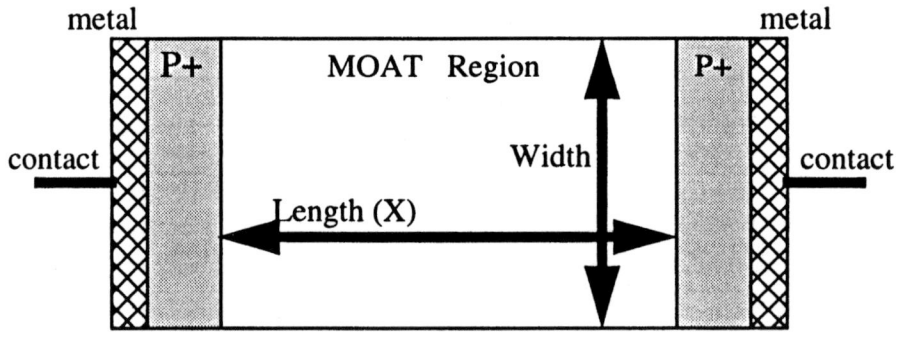

FIGURE 10.4 Top surface view of moats of Figure 10.3.

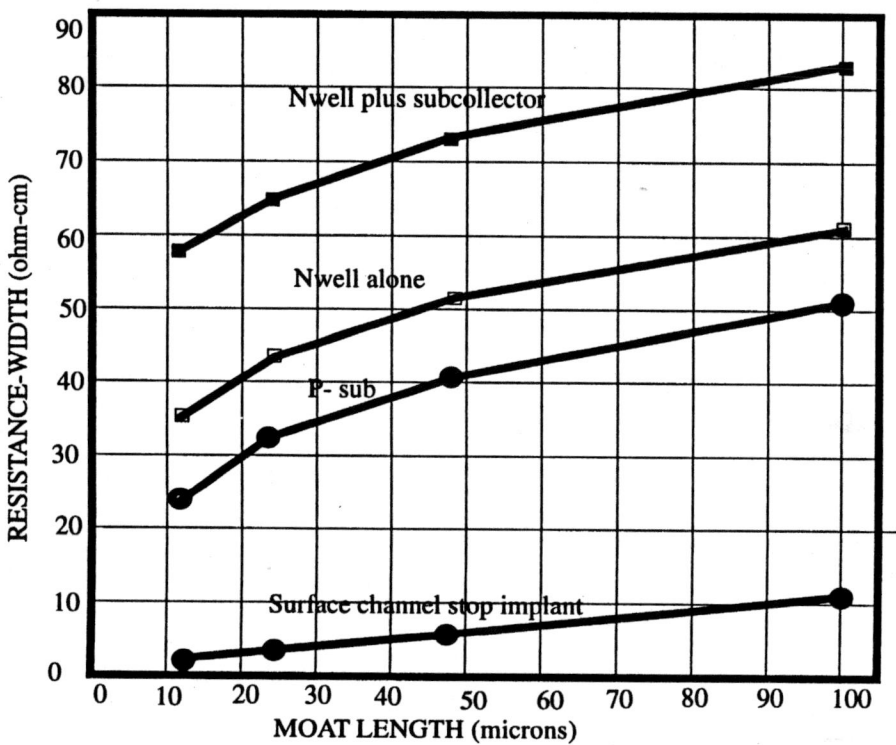

FIGURE 10.5 Simulated Moat Isolation Resistance with different structures.

Because of this shallow lateral current flow near the surface, shallow nwell guard rings are effective at diverting current flow downward into the bulk material and away from protected devices. Likewise shallow surface P+ guard rings can better control local substrate potential around devices due to the non vertical current flow and higher resistance of the bulk substrate. Shielding from shallow surface P+ contact guard rings is dramatically improved by almost an order of magnitude relative to the P+ bulk/P- epi substrates.[10.3]

Controlling Substrate Coupling in Bulk P- Wafers

The coupled noise between resistively separated switching and nonswitching power domains decreases almost linearly with the separation resistance. However, because each domain shorts the substrate within via ties distributed on its ground bus, not much benefit is obtained by moving a circuit from the edge to the center of a power domain; much the same as with P+bulk/P- epi substrates. The improvement of devices in a well over devices in the bulk substrate (pfet vs nfet in nwell process) is not as pronounced due to the shallow lateral current flow. When multiple separated power domains are used it is very difficult to predict the resistances between each of the power domains without detailed simulations of the chip geometries.

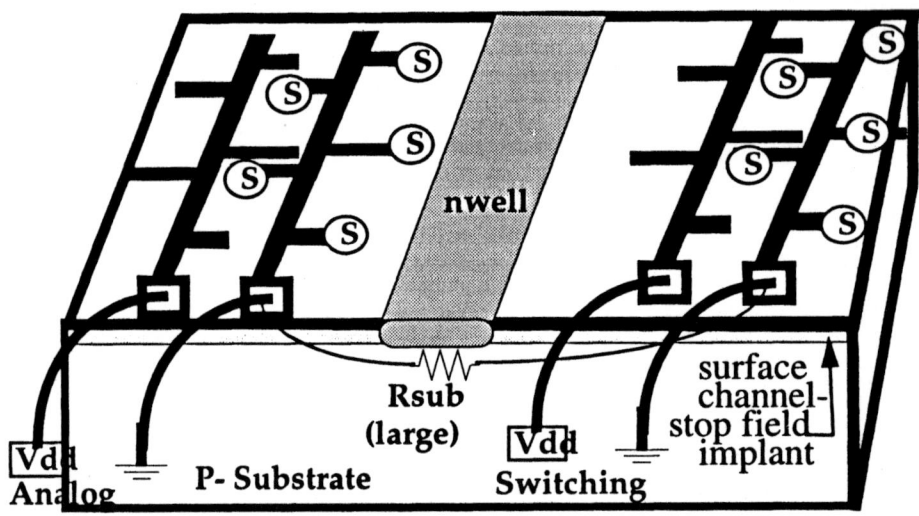

surface distributed substrate contacts are required to prevent latchup

FIGURE 10.6 Substrate Contact on all Power Carrying Grounds.

Bulk P- Wafer Characteristics

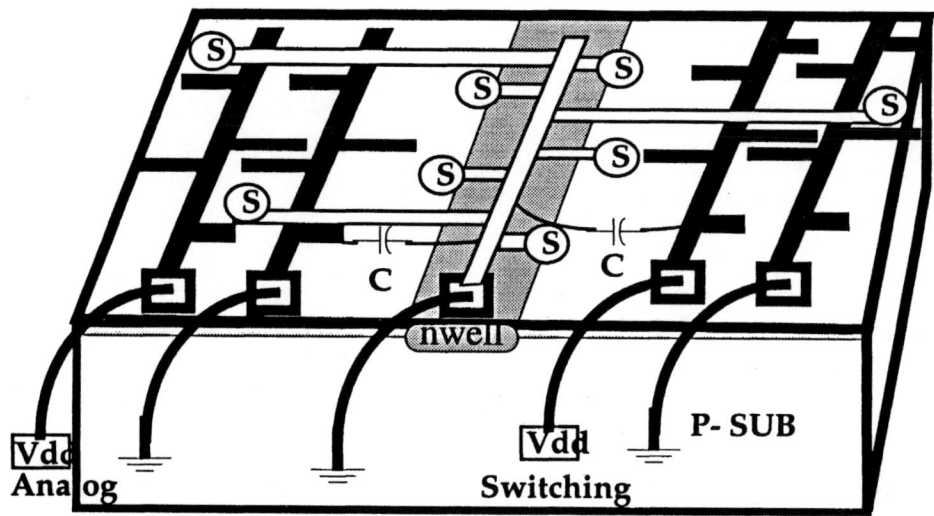

single Kelvin substrate tie ground effectively shorts isolation nwell!

FIGURE 10.7 Substrate Contacts on Kelvin Ground.

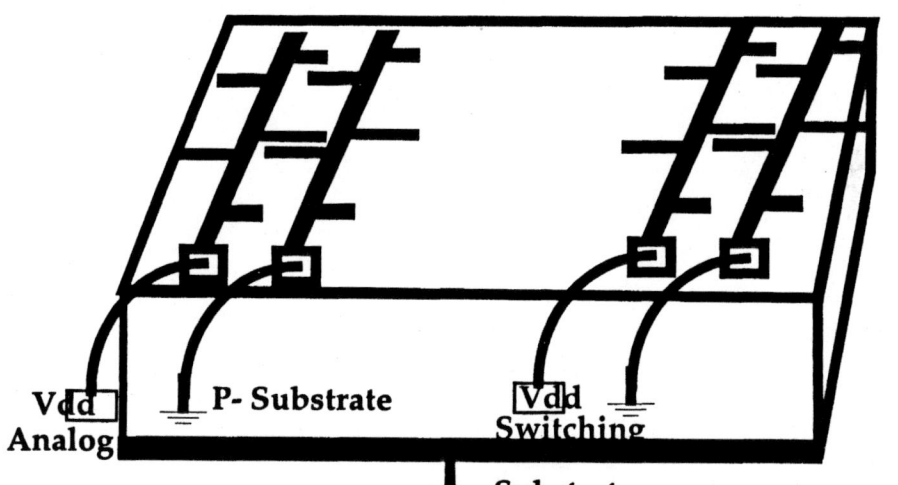

To prevent latch-up wafer should be thinned or higher doped.

FIGURE 10.8 Substrate Contact on Backside Contact.

10.2 Substrate Attenuation Structures

Figure 10.10 shows a technique that combines separation of switching from non-switching power domains and a method to attenuate coupling between the two domains. The situation is depicted for a bulk P- substrate. The separation or moat region can be built using a nwell or other means. In the moat separation region P+ diffusions connected to metal strips and separate bond wires are used to provide attenuation in the isolation or moat area.

Figure 10.9 shows the simplified electrical equivalent circuit. Zan is the equivalent impedance of the non-switching or quiet bus connections to substrate with the equivalent chip/package capacitance and inductance. Zn is the equivalent impedance of the switching or noisy bus connections to substrate with the equivalent chip/package capacitance and inductance. Rs1, Rs2, and Rs3 are the near-surface substrate resistances of the three separate moat isolation regions. Rc1 and Rc2 are the contact and metal resistances of the P+ regions that break the moat isolation area into three pieces. Likewise L1 and L2 are the inductances associated with the two contacts from bond wires and package. Rsd1,2,3 are the deep shunting resistances from node A to node D due to the current flowing deeper in the substrate than the P+ contact diffusions. Rcd1,2 account for the poorer effectiveness of the contacts at points B and C in collecting the deeper substrate currents.

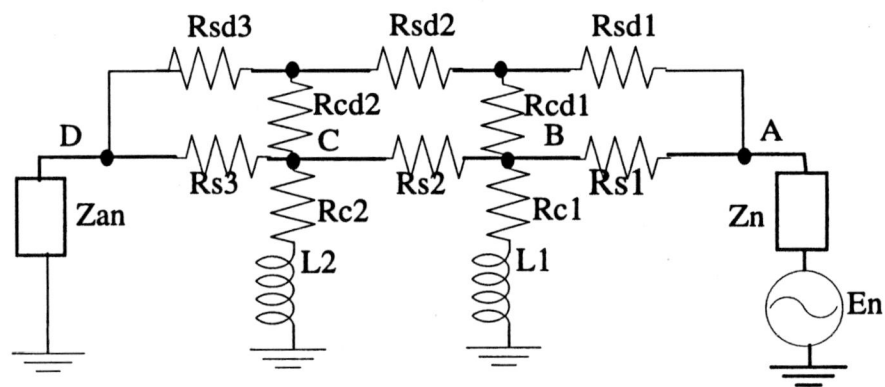

FIGURE 10.9 Simplified electrical equivalent model for substrate attenuator.

Substrate Attenuation Structures

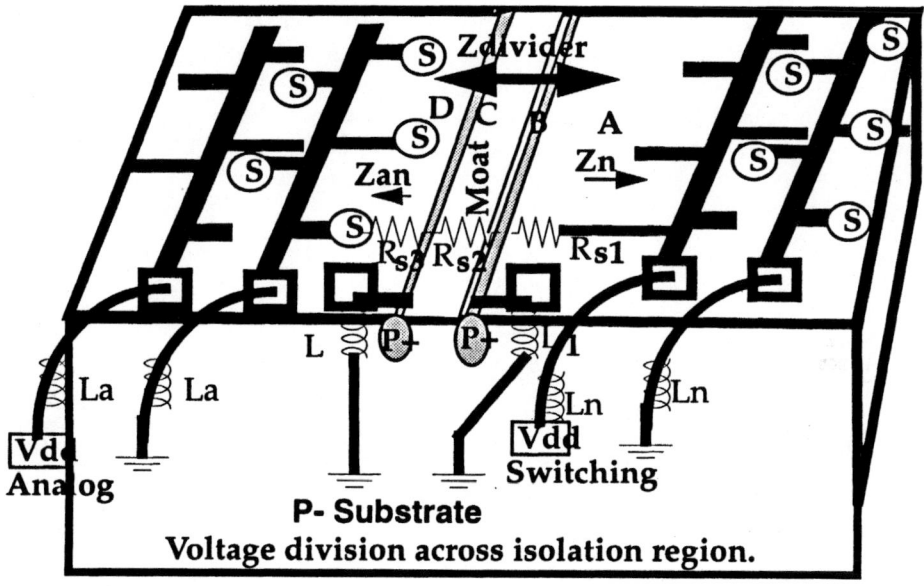

FIGURE 10.10 Substrate Splitting (voltage divider across substrate).

The circuit clearly provides attenuation of the noise at point A before reaching point D with low enough values for L1, L2, Rc1, Rc2 and high enough values for Rs1, Rs2, Rs3. If Rs1,2,3, are large and Rc1,2 and L1,2 are small the attenuation is large. Unshown shunt capacitance across Rs1,2,3 inherent in the isolation moats and due to crossing signal wires limits the usable frequency range of this technique

Controlling Substrate Coupling in Bulk P- Wafers

Figure 10.11 shows the same attenuation structure in a process with P+ buried layers to reduce latch-up. The presence of the P+ buried layers causes the current to be more vertical and go deeper into the substrate. This reduces the effectiveness of the P+ attenuation strips in collecting current crossing the isolation region since the current is now not concentrated at the surface. The effect in the model of Figure 10.9 is to increase the values of Rcd1 and Rcd2 and provide an additional shunting resistance from node A to node D. If the surface P+ contact implants reach through to touch the buried P+ regions then the effect of the buried P+ layers actually helps the substrate noise attenuation. To make this effective the buried layers directly under the the P+ surface contacts must be broken from the P+ buried layers under the analog and switching power domains.

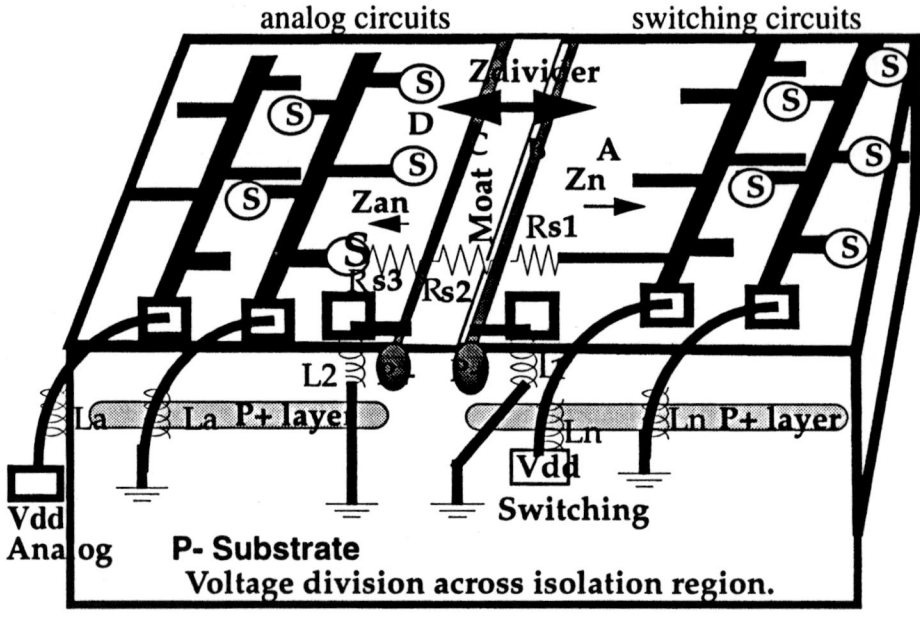

FIGURE 10.11 Substrate Splitting (less benefit with p+ buried layer).

Substrate Attenuation Structures

Figure 10.12 shows how to obtain very high isolation using multiple wells. In this example quiet and switching functions are placed in their own nwell. Each nwell also has dual well capability. Trench isolation is then used between wells to guarantee low capacitance between wells. This process also happens to be radiation hard as well. It has been used to produce less than 1 nano-volt per root hz preamps on the same die with large portions of digital signal processing logic. It was developed at Harris Semiconductor. This is an elaborate process but demonstrates what is necessary to achieve this performance.

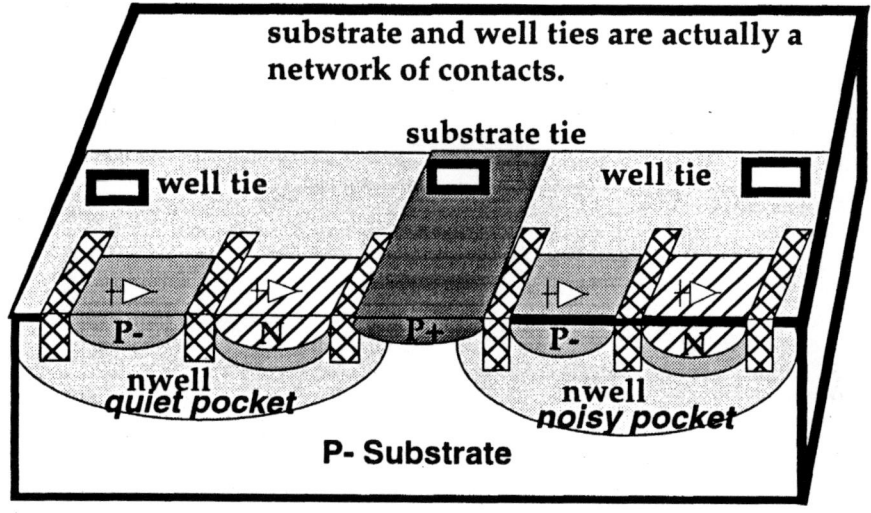

FIGURE 10.12 Multiple substrate well isolation.

Figure 10.13 shows isolation capability available with SOI or Silicon On Insulator technology. SOI, Silicon on insulator processes, have the potential for almost separate chip levels of channel isolation between functions on chip. The switching and non-switching functions can be partitioned in dielectrically isolated islands on the semi-insulating substrates. This technique can deliver the best isolation possible on a common chip substrate. Note that each island must be tied to a potential to collect the capacitive currents injected across the dielectric trench. SIMOX SOI wafers where oxygen is implanted below the silicon surface to produce a layer or oxide under a thin uniform silicon layer are achieving commercial success today. The volume is quite low today but as the defect concentration in the surface layer is brought lower this will be a very fast growing segment Most of the noise problems discussed in this book will be reduced by orders of magnitude for SOI wafers. [10.19]

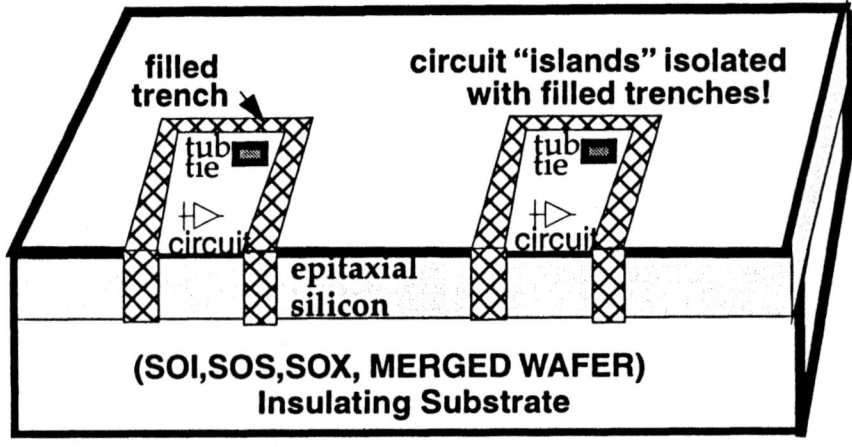

FIGURE 10.13 IDEAL SITUATION IS AN INSULATING SUBSTRATE.

10.2.1 ESD Considerations

Note that P+ substrates provide a common node to dump ESD energy. That way, every pad is protected to every other pad. If you use a P-Minus wafer, you would have to find an ESD path from every pad to every other pad. The problem becomes more complicated as multiple independent power supplies are used to partition switching from nonswitching functions. The ESD path can be provided via parallel opposite polarity diodes between separated supply rails or via a diffused P+ ring at the chip edge that all ESD devices tie to. Each of these also provides another coupling headache between chip pads.

Summary

10.2.2 Potential for Noise Cancellation

The noise coupled to, for example, an A/D converter input can be cancelled if the noise has a fixed amplitude at sampling instants. This cancellation is usually only good at one power supply, temperature, and clock frequency with a fixed amount of switching activity since the amplitude usually varies with all of these variables. This includes noise from other logic circuits clocked from the same sampling clock or higher frequency derivative of the ADC sample clock. Lower frequency divisions of the ADC sample clock will not produce equal amplitude noise pertibations at each ADC sampling instant. At each sampling clock instant the power RLC resonant circuit is stimulated and rings perhaps longer than the sampling cycle. If only a single base frequency clock stimulates the power RLC resonance the resonant coupled noise will eventually reach steady state at the sampling times. This allows the RLC resonant noise to be zeroed or cancelled out as well. If more than one base frequency clock is stimulating the RLC resonant power circuit the RLC noise will not reach steady state with respect to any one of the base clocks and so cannot be fully cancelled out. Note also that the power RLC resonance frequency mixes with the CMOS inverter frequencies since the power fluctuation modulates the current in the switching CMOS devices.

10.3 Summary

This chapter introduced the characteristics specific to a bulk P- or (N-) wafer process. It described several moat isolation characteristics and substrate attenuation structures. Techniques and strategies for minimizing coupling through the chip substrate on mixed-signal ICs using such bulk proceses were introduced.

REFERENCES

[10.1] Timothy Schmerbeck, "Mechanisms and Effects of Noise Coupling in Mixed Signal ICs," EPFL, Switzerland course presentation, June 29-July 10, 1992.

[10.2] T. Schmerbeck, et al., "A 27 Mhz Mixed-Signal Magnetic Recording Channel DSP Using PRML," *Technical Digest of IEEE ISSCC*, pp. 136-137, Feb. 1991.

[10.3] D.K. Su, M.J. Loinaz, S. Masui and B.A. Wooley, "Experimental Results and Modeling Techniques for Substrate Noise in Mixed-Signal Integrated Circuits," *IEEE Journal of Solid State Circuits,* Vol. 28, No.4, pp. 420-430, April 1993.

[10.4] L. D. Smith, *et al.*, "A CMOS-Based Analog Standard Cell Product Family," *IEEE Journal of Solid-State Circuits*, Vol. 24, No. 2, pp. 370-379, April 1989.

[10.5] L. D. Smith, T. Schmerbeck, et al., "A CMOS-Based Analog Standard Cell," *IEEE Journal of Solid-State Circuits*, vol. 24, No. 2, pp. April 1989.

[10.6] H. B. Bakoglu, Circuits, Interconnections, and Packaging for VLSI, Addison-Wesley, Reading, Massachusetts, 1990

[10.7] Charles S. Walker, Capacitance, Inductance and Crosstalk Analysis, Artech House, Boston, 1990.

[10.8] Notes for the IEEE SSCTC workshop on NOISE in MIXED A/D ICs, held in Williamsburg Va, Sept 6-7, 1990. Tim Schmerbeck and Larry Smith. There were no proceedings,

[10.9] T. Gabara, "Reduced Ground Bounce and Improved Latch-Up Through Substrate Conduction," *IEEE Journal of Solid-State Circuits*, 23, No. 5, pp. 1224-1232, October 1988.

[10.10] Henry W. Ott, "High Speed Digital Design", Electromagnetic Compatibility Seminar, Henry Ott Consultants, January 1992.

[10.11] Dr. Tom Van Doren, "Grounding and Shielding Electronic Systems," NTU Satellite Network, February 1991.

[10.12] R. Philpott, T. Schmerbeck, et al., "A 7Mbyte/sec (65Mhz), Mixed Signal Magnetic Recording Channel DSP Using Partial Response Signalling with Maximum Likelihood Detection," *IEEE Custom Integrated Circuits Conference*, May 1993.

[10.13] N. Verghese, D. Allstot, and S. Masui, "Rapid Simulation of Substrate Coupling Effects in Mixed-mode IC's," In *Proceedings IEEE Custom Integrated Circuits Conference,* pp. 18.3.1-18.3.4, May 1993.

Summary

[10.14] T. Schmerbeck, "Design Strategies for Reducing the Effects of Noise Coupling in Analog and Mixed-Mode ICs," in presentation for course on *Practical Aspects of Analog and Mixed-Mode IC Design*, Beaverton Oregon, May 18, 1993.

[10.15] B. R. Stanisic, R. A. Rutenbar, and L. R. Carley, "Power Distribution Synthesis for Analog and Mixed-Signal ASICs in RAIL," In *Proceedings IEEE Custom Integrated Circuits Conference,* pp. 17.4.1 - 17.4.5, May 1993.

[10.16] Internal IBM Technical Report by Larry Smith, IBM, San Jose, CA.

[10.17] Internal IBM Technical Report by Tim Schmerbeck, IBM, Rochester, MN.

[10.18] Kazuo Kato, Hideo Sato, Yasuji Kamata, Kenkichi Yamashita, and Seiichi Ueda, "A 300-MHz Monolithic Video Current Driver for High-Resolution CRT Applications," *IEEE Journal of Solid State Circuits,* Vol. 24, No.4, pp. 1110-1117, August 1989.

[10.19] D. Flandre, J. P. Colinge, "Status and Trends of SOI," In *Proceedings IEEE European Solid-State Circuits Conference,* Ulm, Germany, September 20-22, 1994, pp. 18-27.

CHAPTER 11
Chip/Package Shielding and Good Circuit Design Practice

11.1 Far Field Radiated Emissions

It is now occurring that radiated emissions from a single packaged IC are exceeding FCC Class B specifications in the United States as well as other emission standards through-out the world. Violation of radiated emission specifications will usually occur well before the emissions start to functionally perturb the design itself. For radiated Class B designs that have clock frequencies exceeding 100Mhz the specification on the emissions envelope at 5 meters distance from the device extends to 1Ghz in frequency. Class A specifications are taken at 10 meters distance from the radiating device. The usual definition of far field radiated noise is that the distance from source $> \frac{\lambda}{2\Pi}$. Antenna length, $L \sim > \lambda$ (wavelength) for *an EFFICIENT* antenna. A "rule of thumb" to minimize radiated emissions is to keep the chip package linear dimensions $L < \frac{\lambda}{20}$ and to keep the chip/package $Area < \frac{\lambda^2}{800}$. An example:
$\lambda = 30cm$ @f=1Ghz, $\frac{\lambda}{2\Pi} \approx 4.8cm$, $\frac{\lambda}{20} \approx 1.5cm$ $\frac{\lambda^2}{800} \approx 1.125cm^2$.

Note that the 68 pin PLCC package has maximum linear dimensions along a diagonal of the package of 3.56cm and maximum area of 6.3 cm squared. This rule of thumb indicates that this package is very capable of exceeding radiated emissions specifications at 1Ghz given the proper excitation from the IC. Table 11.1 shows a chart of wavelength vs frequency in air with the shaded rows showing possible chip/package dimensions that come within 1/20th of the wavelength.

TABLE 11.1 Linear Dimensions vs Frequency in Air.

Frequency	Wavelength	Wavelength/20	Wavelength^2/800
1Mhz	300 meters	15 meters	112.5 sq m
10Mhz	30 meters	1.5 meters	1.125 sq m
100 Mhz	3 meters	15 cm	112.5 sq cm
500 Mhz	60 cm	30 mm	4.5 sq cm
1 Ghz	30 cm	15 mm	1.125 sq cm
10 Ghz	3 cm	1.5 mm	1.125 sq mm
100 Ghz	3 mm	150 microns	.01125 sq mm

Consider that for a radiated EM plane wave in free space $\frac{E}{H} = Z_0 = 377\Omega$. Neither E or H dominates which is the definition of a plane wave. By looking at the rate of change of voltage and current with respect to time in a circuit or IC an estimate can be made of whether the radiation will be dominated by the magnetic field (inductive coupling) or electric field (capacitive coupling). If (dv/dt)/(di/dt) <<377 ohms in a circuit or IC the radiation will be dominated by magnetic or inductive fields. If (dv/dt)/(di/dt) >>377 ohms in a circuit or IC the radiation will be dominated by electric or capacitive fields. Here dv/dt and di/dt are the fastest voltage and current transients occurring in the circuit or IC. Since there will be multiple radiators in the circuit or IC the area of each antenna will also become a factor.

11.1.1 Differential Mode Radiation due to a magnetic dipole

Differential mode radiation due to a magnetic dipole (current loop) has an emissions envelope of its electric field given by $E \propto F^2 A I c$. E is the electric field, F is the frequency of the emission, A is the loop area, I is the current in the loop and c is a constant. The electric field is used for a reference and the H field can be determined from it using Maxwell's equations. Assume that the current excitation waveform in the current loop is a square wave. For finite rise time, 50% duty cycle, square wave only odd harmonics are ideally present and 64% of energy is in the fundamental square wave frequency. Because the square wave excitation amplitude drops off with increasing harmonics but the dipole radiated emission increases with the square of the frequency we find that the peak of the radiated field increases from the fundamental to the fre-

Far Field Radiated Emissions

quency corresponding to the square wave rise time at 20db/dec and then flattens out. So the emissions from a square wave excitation of a current loop will have an emissions envelope that extends theoretically to infinite frequency. For this reason it is usually differential mode radiation that causes the most problems in high frequency design. Figure 11.2 shows this pictorially. Since differential mode radiation involves signal currents it is possible to reduce this radiation by reducing the magnitude, rise time, and or frequency of the current waveform. By reducing the loop area of the current loop the radiation can also be reduced. If the loop net spacing to a nearby ground lead is less than the spacing to a ground plane the signal will return on the smaller loop. If there is such a ground lead placed on either side of the loop, the currents will split from the loop and be out of phase to partially cancel the radiated wave.

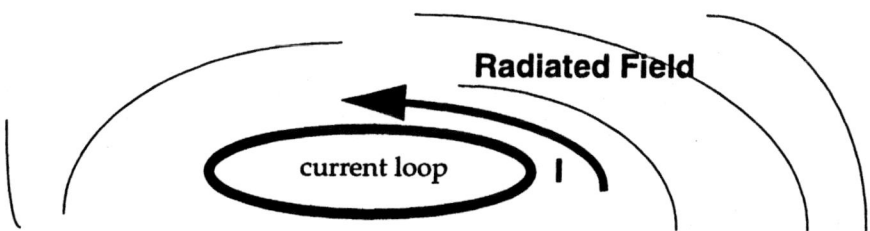

FIGURE 11.1 Radiated field from a magnetic dipole or current loop.

Chip/Package Shielding and Good Circuit Design Practice

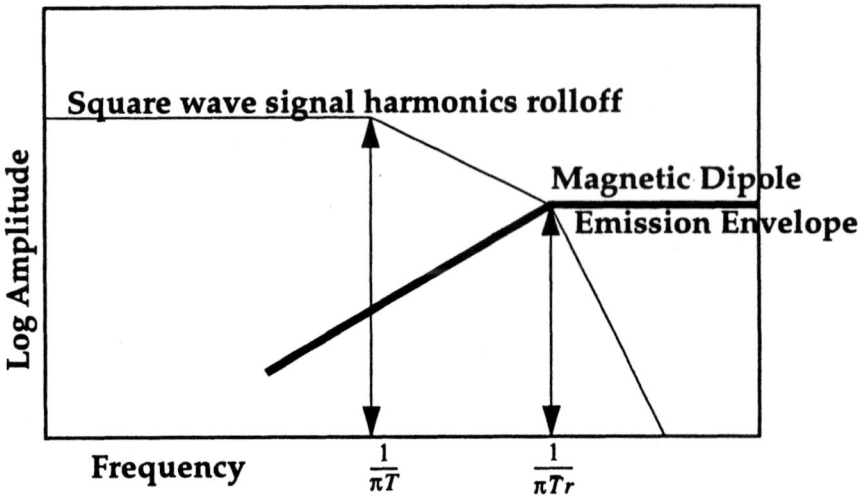

FIGURE 11.2 Magnetic dipole radiated emissions envelope from square wave source.

11.1.2 Common mode excitation of interconnecting leads acting as an electric dipole

In contrast to differential mode radiation, common mode radiation usually does not involve signal currents. Figure 11.3 shows an example of the signal differential across the chip substrate producing a common mode signal that is radiated through the package leads. This situation is essentially a monopole over a ground plane. Common mode radiation due to a monopole or dipole has an emissions envelope of its electric field given by $E \propto FLI_{cm}c$, where E is the electric field emissions envelope, F is the frequency of the emission, L is the antenna length, Icm is the common mode current, and c is a constant. The loop shown in Figure 11.3 is usually completed in free space. The radiation emissions envelope from a square wave current excitation is constant from the square wave fundamental till the rise time frequency and then drops off at 20db/decade. For this reason, common mode radiation is usually not a problem at very high frequencies relative to the fundamental. Figure 11.4 shows the relationship pictorially. To reduce common mode radiation you need to reduce L, Icm, common mode potential (substrate differntial noise between power domains), signal rise time, signal frequency, or use sine wave source excitation instead of a square wave. A smaller IC package will reduce L.

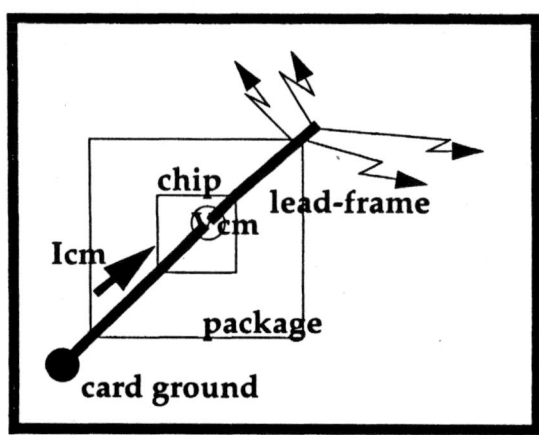 

Equivalent Circuit: monopole over a ground plane.

FIGURE 11.3 Common Mode Radiation from a package.

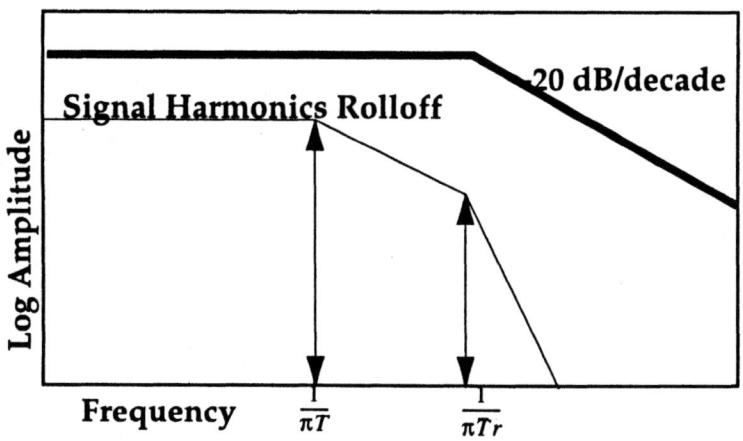

FIGURE 11.4 Electric Dipole Radiation Envelope.

Chip/Package Shielding and Good Circuit Design Practice

11.1.3 Radiated, then received EM waves

This is usually not a problem on IC wires themselves until tens of Ghz. It can be a problem on long package leads and especially for sensitive radio or optical receivers. The sensitivity depends on the gain of the receiver and the strength of the transmitter, Antennas where $L_{antenna} < \frac{\lambda}{100}$ are very poor and inefficient antennas. Note however that a car AM antenna has $L_{antenna} < \frac{\lambda}{200}$ but still receives well due to a 5KW transmitter power at distances of 20 miles. A strong transmitter negates the need for an efficient antenna.

11.2 Effect of Chip Signal Isolation/Shielding Techniques on Noise

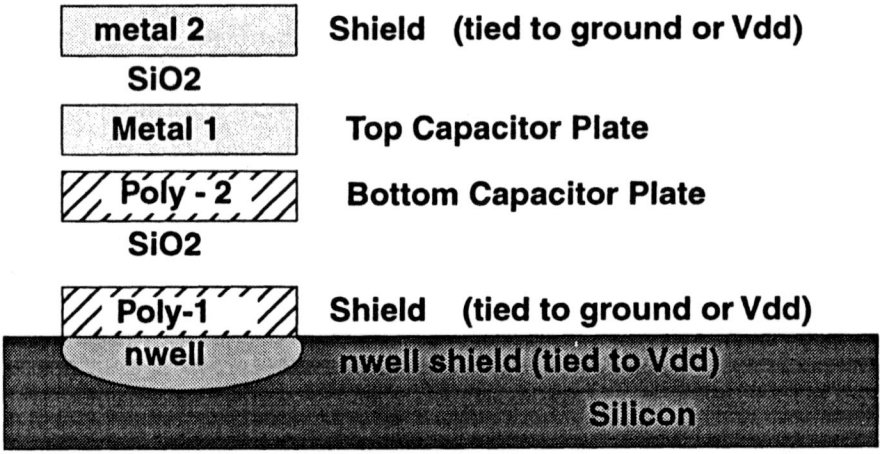

FIGURE 11.5 Faraday shielding of circuit nodes.

Figure 11.5 shows a metal-1 to poly-2 capacitor laid out with a metal-2 faraday shield above and a poly-1 and nwell shield below. Shield conductors are also possible at the sides of the capacitor as discussed earlier. Shields can become contaminated with noise making them efficient noise couplers to the circuit to be shielded. It is often better to replace a large higher impedance shield with a separation distance. The separation attenuates signals and cannot become contaminated. Shielding becomes more difficult for devices diffused into the substrate or epitaxial layer. Figure 11.6 and Figure 11.7 show various simplified shield structures for a P- bulk and P+ bulk/P- epitaxy substrate respectively. In Figure 11.6 the current flow is concentrated near the surface and the surface P+, N+, and trench guard rings are not dramatically worse in shielding than the wells which protect the entire bottom surface of the circuit being shielded. The closer the guard ring is to the device being shielded the better the shield. In Figure 11.7 the presence of the P+ lower substrate causes most of the current to flow vertically. In this situation the wells give dramatically better shielding than the guard rings. The guard rings are virtually ineffective at shielding from noise currents from the bulk P+ substrate. The guard rings are still effective at reducing current injection from an adjacent device or circuit since much of that current can still flow for short distances in the field stop implant regions while it is spreading out toward the substrate.

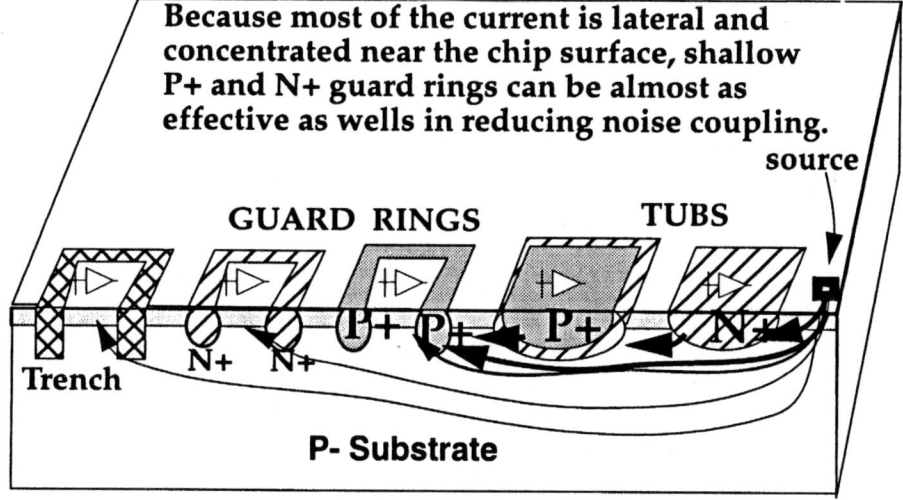

FIGURE 11.6 Guard rings and well isolation in P- Bulk substrates.

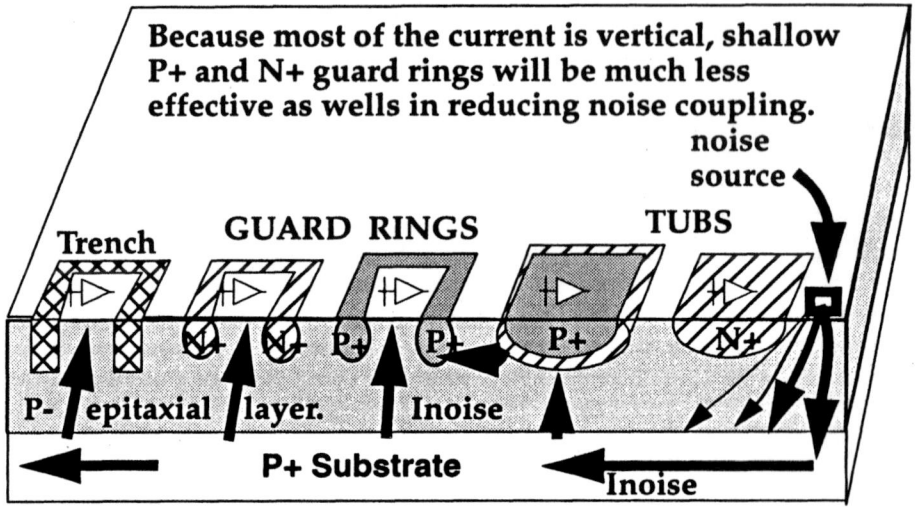

FIGURE 11.7 Guard rings and well isolation in P+ Bulk/P- epitaxial substrates.

11.2.1 Nwell structure

An Nwell Structure protects analog circuits from free carriers and substrate noise and provides decoupling capacitance for the power supply. Figure 11.8 shows an N-well structure at the end of the analog book design area and the wiring channels in the analog terrain. Nwells reside below thick oxide so wiring capacitance is not affected. The Nwell is connected to Vdd using the first-level metal power redistribution straps of the analog power bus structure. Stray holes and electrons are injected into the substrate by logic and analog books and can be considered to be carrier noise. The Nwell Structure forms an effective carrier barrier to free holes and electrons. As shown in the cross section drawing, the Nwell is 4.5 um deep. A highly doped substrate is 8 um deep or 3.5um below the Nwell. The Nwell is biased at Vdd and forms an effective collector of electrons. Holes are contained by the Nwell structure or forced down into the highly doped substrate. Stray electrons and holes are thus prevented from crossing design rows. Nwell straps, shown are used to exclude carriers from analog books that are particularly sensitive to carriers or contain carriers in books that are known to emit them. The Nwell structure provides a safe area for substrate contacts. Once holes are forced down into the highly doped substrate by the carrier barrier, they will tend to stay there until they are attracted up by an ohmic contact at ground potential in the P-

Effect of Chip Signal Isolation/Shielding Techniques on Noise

epi. Substrate contacts are avoided in analog books because there is an IR voltage drop associated with current that flows out of the highly doped substrate through the epi to a substrate contact. The altered potential at the epi surface may affect the threshold voltage of an FET that is intended to match another FET or could induce latchup in an extreme case. As shown the substrate contact is completely surrounded by the Nwell structure biased at Vdd and is therefor a safe contact. Voltage drop in the epi on the way to the substrate contact will have no detrimental effect on analog books or pose any latchup exposure. The resistance from the contact to the highly doped substrate has been calculated to be 1 Kohm. With 18 contacts per double row and 10 analog double rows, the resistance from the analog power bus to the highly doped substrate is 5.6 ohms. The highly doped substrate is 50 milli-ohms per square and can be considered to be a constant potential surface.

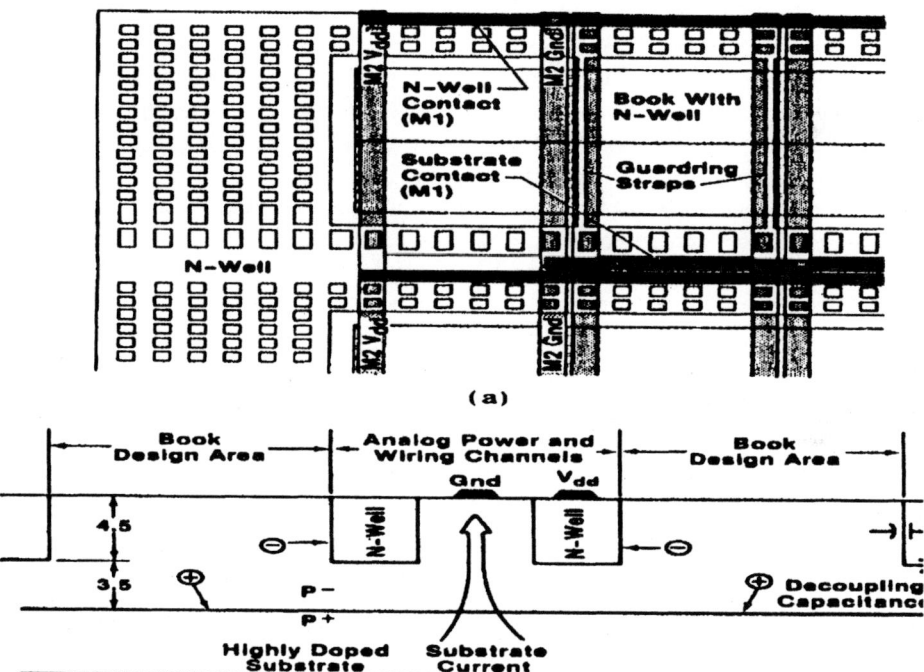

FIGURE 11.8 Nwell structure for carrier barrier, substrate contact, decoupling C.

The Nwell structure forms an effective decoupling capacitor from Vdd to GND. Both the bottom surface of the Nwell and the sidewalls contribute capacitance. The holes in

Chip/Package Shielding and Good Circuit Design Practice

Chip/Package Shielding and Good Circuit Design Practice

the Nwell increase the sidewall area and therefor the "capacitance density." With the Nwell structure shown, 15 pF is gained for each design row. This gives 150 pF for an analog terrain that is 10 rows high. The capacitance is distributed very close to analog books which is the most effective position possible.

11.3 Effect of Packaging on Noise

We have already considered the effects of package capacitance and coupling for a simple package example. Packages with ground planes can also reduce radiated emissions and line to line coupling. However, packages with any form of floating metal can dramatically increase radiated emissions as well as chip crosstalk. Floating metal increases capacitive effects and increases electromagnetic effects. An example of this problem is a high gain amplifier on an IC. Since stray output to input coupling is undesirable and causes spurious oscillations the inputs and outputs are kept widely separated. A floating piece of metal in the package can couple these together through the capacitance of bond wires or lead frame leads that run near the floating metal. Common sources of package floating metal include, package lids, the chip leadframe heat spreader or die attache pad, or the package itself. Capacitive currents coupled to the floating metal often find a return path to ground via radiation. Figure 11.9 shows a cross section of a quad flat pack package that is made from two machined metal halves. In this particular package, the bottom metal of the package is close enough to the card ground plane that 80pF of capacitance exists from the bottom half to the card ground. This bottom metal half is not really floating due to the large capacitance to ground. Each lead of the leadframe has only a fraction of a pF of capacitance to the bottom or top half of the package. The metal top half of the package is also capacitively coupled to the bottom half of the package. The approximate capacitance is 30pF. The chip die attache area under the chip is also capacitively coupled to the bottom half by approximately 80pF as well. There is not a dc connection from the die attache pad to the metal package bottom. By bonding from a leadframe pin, that is connected to ground, to the chip die paddle a parallel capacitive connection is made from the package bottom to card ground. This can be very beneficial when the package is in a socket and the capacitance from the package bottom to card ground increases dramatically due to the further separation from the card ground plane. It is also possible to use the top and bottom metal halves of the MQUAD package as faraday shields. Their significant capacitance to ground makes them pseudo ground shields at high enough frequencies that their capacitance looks low impedance. Any currents flowing within the package will be shielded by this effect. Earlier we talked about substrate separation between switching and non switching function to achieve

Effect of Packaging on Noise

something like 50 ohms separation. There are still switching signals that must communicate between the two switching domains. Since the return path on chip of 50 ohms is greater impedance than the outside the package return path, the current will return via an outside the package loop. This increased loop area will radiate more noise. Often it is possible to link the two separated switching and non-switching grounds via a common bond from each to the chip die attache paddle. This forces the currents to flow inside the MQUAD package for faraday shielding and also reduces the radiation loop area. The latter affect of reducing the radiation loop area will also show radiated emission reductions even in a plastic or non-conductive package.

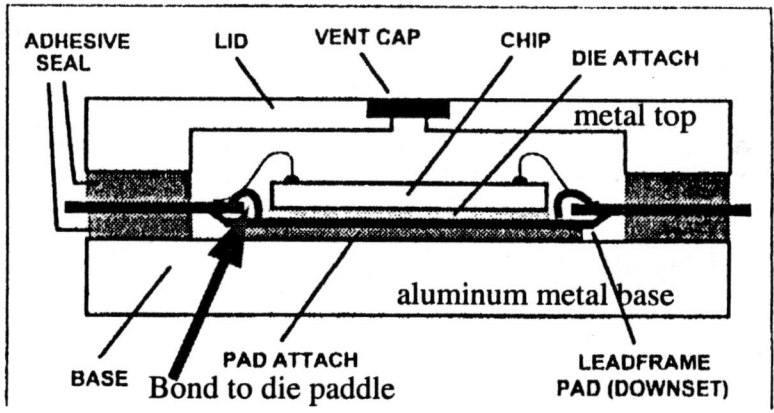

FIGURE 11.9 Metal Quad Flat Pack or MQUAD Package. (Courtesy of OLIN).

In a normal plastic package, the top and bottom portions of the package are injection molded plastic. In this situation it is possible to float the metal die attache area under the chip by not bonding to it. Now leadframe wires will be capacitively coupled to the floating metal die attache area. Because the die attache area is under the IC it readily couples signals from one side of the chip to the other thru the capacitance to the die attache area. Bonding from leadframe ground pins to the die attache area reduces the coupling dramatically and also reduces radiation from the floating metal. Figure 11.10 shows the many more possible chances of coupling and interaction possible in a multilayer, multiconductor, multi-chip module. Figure 11.11 and Figure 11.12 show some fairly low cost methods of achieving multi-chip modules with a 8-pin dual in line package and a quad flatpack. Both of these approaches form the leadframe in such a way as to produce two separate chip die attache paddles. This two chip approach can

Chip/Package Shielding and Good Circuit Design Practice

obviously produce a much high impedance separation between a switching and non-switching chip. The chip to chip communication lines can again be contained inside the package along with ac return current paths to reduce radiated emissions levels.

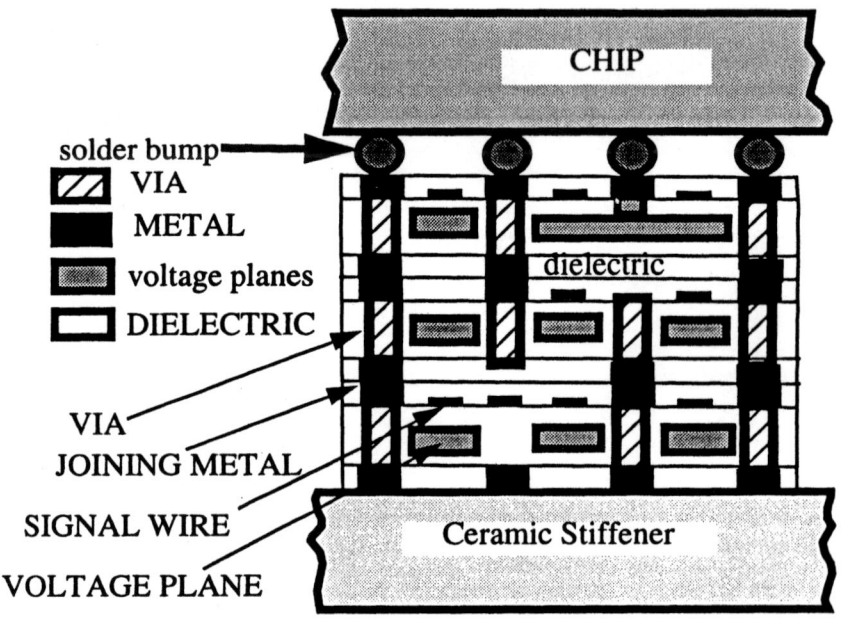

FIGURE 11.10 Cross-section of a multi-layer, multi-plane, package.

Effect of Packaging on Noise

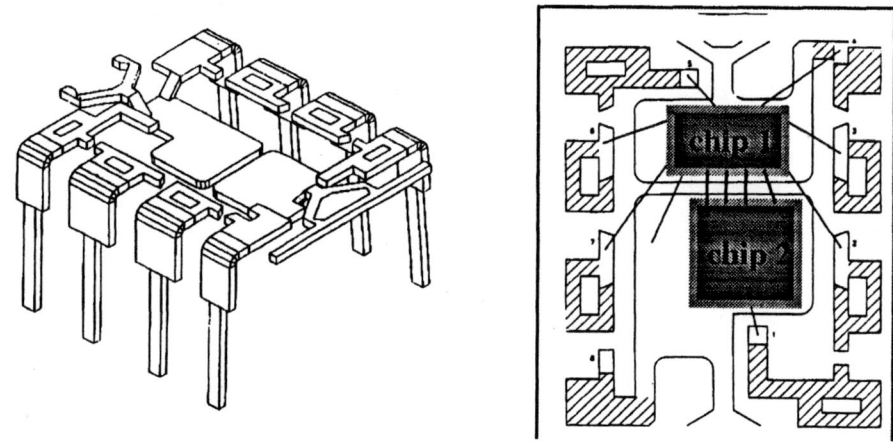

FIGURE 11.11 Multi-chip module with 2 chips on common leadframe (8 PIN DIP).

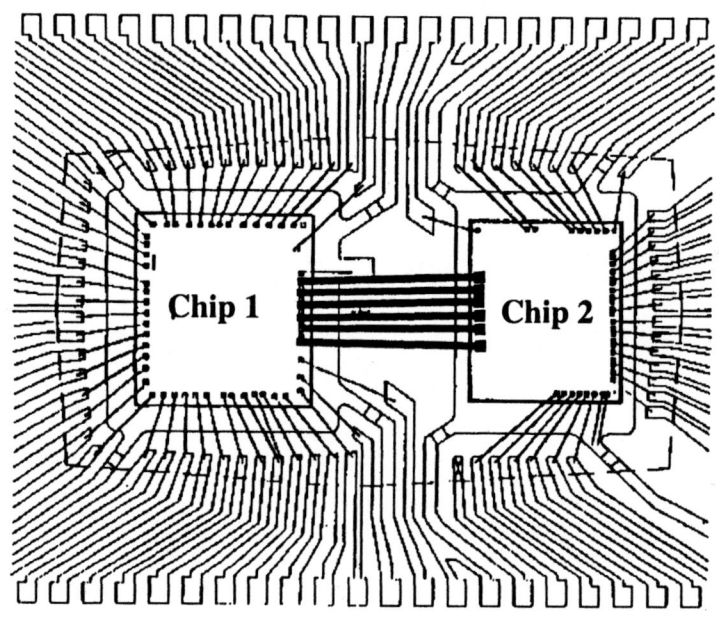

FIGURE 11.12 Multi-chip module with 2 chip on common leadframe. (QFPK).

Chip/Package Shielding and Good Circuit Design Practice

11.4 Effect of Card Layout and Referencing on Noise

The card or carrier that the packed IC is connected to plays a key role in obtaining a minimum noise environment. On of the key factors is card level decoupling capacitance. The series resonance of the capacitance system must be taken into account when designing the decoupling system. Generally for a switching waveform, the on chip capacitances provide the current for decoupling during the rise and fall times, the card ceramic (or high frequency) capacitors recharge the chip capacitors between edges over a few cycles and the card bulk capacitors (typically tantalum) recharge the ceramic capacitors over many cycles of the waveform. A failure in sizing the capacitors or in frequency response will result in an increased drop in the supply voltage due to switching transients. Although there are recommendations in this chapter that support the splitting of the grounds and supplies between switching and non-switching supplies on-chip, this is usually not recommended for the card level. Splitting the card or ground planes creates discontinuities in driver return currents and increases the loop area of these return loops. Both of these effects increase the level of radiated emissions from the card. The increased area of the return loops on card over the chip return loops makes it more dangerous to split the supplies. The chip supplies and ground were split due to the difficulty in obtaining a low enough impedance on either supply rail. Because the impedance was not sufficiently low the supplies with switching activity carried considerable supply bounce and noise. At the card level, low impedance power planes with distributed power decoupling capacitance allow the card supply rails to be low impedance with minimal noise levels. This allows for a single plane for card ground and each supply rail.

11.4.1 Circuit/component grounding

Figure 11.13 shows the difficulties in referencing on-chip signals to off-chip power rails. In part (a) an on-chip current reference is generated using an operational amplifier that forces a fixed on-chip reference voltage across an external resistor Rext. The external resistor is used to obtain some programmability on the current in different applications of the IC. Because the external resistor returns to card ground and the chip reference is referenced to chip ground, any noise between the two grounds appears as differential mode noise on the op-amp inputs and thus noise current on the on-chip current reference. As discussed earlier, the chip ground may carry large resonant bounce and output coupled noise components that the card ground may not have present. In part (b) the external resistor Rext is returned to the chip ground bus through another module pin that is not bonded out to card ground directly. This Kelvin power connection carries only the signal return current of the resistor Rext and

Effect of Circuit Topology on Noise

makes the chip ground noise a common mode noise component at the operational amplifier inputs. The inductance and capacitance of the Rext signal return pin will filter some of the chip ground noise at high frequency so this is not a solution for all frequencies. In general it is better to reference no single ended signals to card ground or card Vdd for the reasons given in the above example. It is possible to differentially reference signals to card ground or Vdd to keep the noise a common mode component.

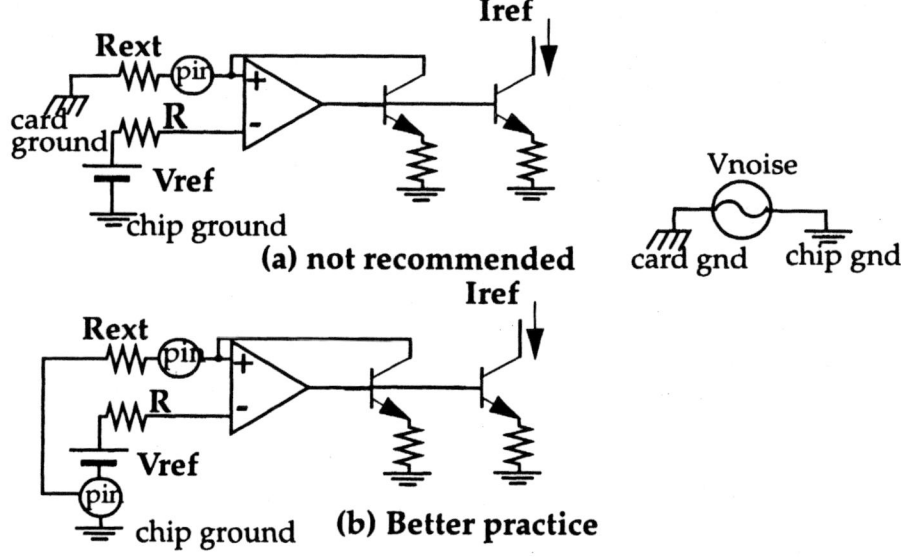

FIGURE 11.13 Difficulties Referencing chip signals off-chip.

11.5 Effect of Circuit Topology on Noise

The effect of circuit topology on coupled noise is a very broad topic for in-depth treatment here. In general it is good to keep circuit impedance levels as low as possible, especially for sensitive circuit nodes to limit coupling from adjacent signals and the chip substrate and power rails. This is a difficult thing to do in low power applications where circuit impedance levels must be high for low power. By the same token it is beneficial to bandwidth limit the circuits to have the minimum required bandwidth.

Chip/Package Shielding and Good Circuit Design Practice

This will limit the circuit sensitivity to broadband noise and possibly eliminate susceptibility to noise in problem frequency ranges. Often the outputs of circuits can be made integrating so that periodic noise such as power supply bounce can be integrated out over many cycles leaving only the small cycle to cycle variations to contend with. The circuit gain and dynamic range should also be limited to what the application needs. Sensitive nodes should be constructed or shielded so as to have minimum capacitance to noise sources such as the chip substrate and power supply rails. In the case of switched capacitor circuits the same design practices that make designs parasitic insensitive tend to limit susceptibility from substrate coupled noise. Limiting switched capacitor circuits so that only one sensitive node exists per amplifier makes it possible to shield these few sensitive circuit nodes. Since the chip may have multiple separated supply rails, the same problems exist on chip as in cards for referencing circuits between separate supply domains. The design of differential configurations will aid in the rejection of common mode supply and substrate noise. At high frequencies the circuit layouts of differential circuits must be exactly symmetrical to achieve common mode rejection. Bias references for circuits such as voltage or current references must be distributed differentially to prevent noise pickup. A single ended voltage referenced to a local chip ground to bias the gate of a MOS current source cannot be translated to a distant point on the ground bus due to dc and ac voltage drops on the power bus as discussed earlier. However, if a number of current sources are produced locally and the currents are distributed to many distant points on the bus and re-referenced locally, the bus drops will not cause an error in the reference.

11.5.1 Summary of Recommended Design practices

The outline below gives a high level overview of the kinds of things that can be done in an IC design to minimize susceptibility to coupled noise.

Noise Prevention Techniques

REDUCE NOISE GENERATION (Quiet the Talker)

Avoid switching large transient supply current:

-Use balanced current steering vs voltage switching internal logic family if possible.

-Avoid CMOS logic families if possible. Alternatives to CMOS "DO" exist such as FSCL.

-Shut down all switching functions or logic/drivers not in use.

Effect of Circuit Topology on Noise

-Ramp clocks or reduce their rise time; also reduces clock feedthrough.

Reduce Chip I/O driver generated noise:

-Use slowest drivers necessary to meet requirements.

-Use controlled voltage rise time I/O drivers.

-Stagger switching of circuits on internal and I/O bus

-Use reduced voltage swing I/O drivers.

-Use balanced current steering drivers to limit di/dt to chip power rails.

-Place large drivers close to power returns to limit loop inductance.

Lower switching function bus impedance:

-Provide gridded, multi-level, or other low resistance logic/driver power bus.

-Provide distributed on-chip decoupling since off-chip capacitors self resonate at much lower frequencies.

-Maximize the number of chip power pads/pins to minimize power inductance.

-Since the CMOS excitation of the power rails is, to first order, out of phase; the substrate current injection can be cancelled with equal capacitance to substrate from the Vdd and ground rails.

-Use one base clock, within a bus or isolation well to avoid beat frequency or mixing problems when possible.

Noise Reduction Techniques

ISOLATE SENSITIVE CIRCUITS (Isolate The Listener)

Provide noise reduction from talker to listener by whatever means possible:

Maximize impedance from noise source to circuit where the noise source can be the chip substrate. The heavier doped the substrate the worse this problem is unless the substrate can be low impedance grounded with a chip backside contact.

Utilize on chip shielding using diffusion, implant, poly, or metal layers.

Provide diffused guard rings and/or wells at minimum distance to the sensitive circuit.

Physically separated package/chip power connections for noisy and sensitive circuits.

Where this is not possible a custom STAR power routing is usually required.

The worst near field coupling occurs at the package leads so I/O assignment is crucial.

Chip/Package Shielding and Good Circuit Design Practice

The routing of chip signal wires must provide strict separation of noisy & quiet wires.

When lightly doped substrates are used physical separation between power domains reduces power bus to power bus coupling through the substrate. N or P wells can be used to provide this physical separation.

Physical separation of circuits on chip.

Logic/noisy signals buffered in analog terrain to remove power bounce noise.

Proper placement of substrate contacts.

Minimize impedance from quiet buses to card ground/Vdd.

Design of circuits for maximum supply and common mode noise rejection.

MAKE THE ANALOG MORE NOISE TOLERANT (Close the Listener's ears)

Design For High CMRR, PSRR.

Use Minimum Required Bandwidth.

Band-limit your signal.

Keep Internal Node Impedances Minimum.

Keep sensitive nodes on-chip.

Differential Circuit Topologies.

Differential Circuit layout.

Hand crafted layout often necessary.

Differential Voltage or Current Distribution.

Balance noise coupling where not differential.

Return All External Programming Components Back To Chip Ground or Vdd Via Kelvin I/Os (Outside of CMRR, PSRR rejection freq range).

11.6 Summary

This chapter surveyed techniques and strategies aimed at minimizing radiated emissions. It also looked at the effects of chip isolation/shielding and packagin on coupled noise. The effects of card layout, referening and circuit topology on noise were also discussed. Finally a summary of good circuit design practice for reducing coupled noise was given. This included both design techniques (noise prevention) and layout techniques (noise reduction).

REFERENCES

[11.1] Timothy Schmerbeck, "Mechanisms and Effects of Noise Coupling in Mixed Signal ICs," EPFL, Switzerland course presentation, June 29-July 10, 1992.

[11.2] T. Schmerbeck, et al., "A 27 Mhz Mixed-Signal Magnetic Recording Channel DSP Using PRML," *Technical Digest of IEEE ISSCC*, pp. 136-137, Feb. 1991.

[11.3] D.K. Su, M.J. Loinaz, S. Masui and B.A. Wooley, "Experimental Results and Modeling Techniques for Substrate Noise in Mixed-Signal Integrated Circuits," *IEEE Journal of Solid State Circuits*, Vol. 28, No.4, pp. 420-430, April 1993.

[11.4] Notes for the IEEE SSCTC workshop on NOISE in MIXED A/D ICs, held in Williamsburg Va, Sept 6-7, 1990. Tim Schmerbeck and Larry Smith. There were no proceedings,

[11.5] Henry W. Ott, "High Speed Digital Design", Electromagnetic Compatibility Seminar, Henry Ott Consultants, January 1992.

[11.6] Dr. Tom Van Doren, "Grounding and Shielding Electronic Systems," NTU Satellite Network, February 1991.

[11.7] T. Schmerbeck, "Design Strategies for Reducing the Effects of Noise Coupling in Analog and Mixed-Mode ICs," in presentation for course on *Practical Aspects of Analog and Mixed-Mode IC Design*, Beaverton Oregon, May 18, 1993.

[11.8] Kazuo Kato, Hideo Sato, Yasuji Kamata, Kenkichi Yamashita, and Seiichi Ueda, "A 300-Mhz Monolithic Video Current Driver for High-Resolution CRT Applications," *IEEE Journal of Solid State Circuits*, Vol. 24, No.4, pp. 1110-1117, August 1989.

CHAPTER 12   *A Design Example*

12.1 Design of a Mixed-Signal IC

Figure 12.1 shows the architecture of the mixed signal IC design of reference [12.10]. The IC is part of the read/write channel of a rigid media disk drive. A 20mV read signal is variably amplified, sampled and equalized under digital control with maximum likelihood sequence detection. An analog AGC (Automatic Gain Control Amplifier), Buffer amplifiers, delay lines, multiple VCOs (Voltage Controlled Oscillator) and PLLs (Phase Locked Loop), D/As and A/D coexist with 6K CMOS gates clocked at 27Mhz. This 5.5mm square chip, packaged in a 68-pin PLCC, uses a 5V, BICMOS process with 1 um Leff and 6 Ghz NPN. The architecture of this IC is explained in reference [12.15]. In Figure 12.1 the amplified signal from the disk surface (A) must be equalized and sampled as shown in waveform (B). Waveform (W) shows the write current waveform that wrote the data of waveform (A) on the disk surface. Figure 12.2 shows a more complex write waveform and the resultant data of waveform (B).

A Design Example

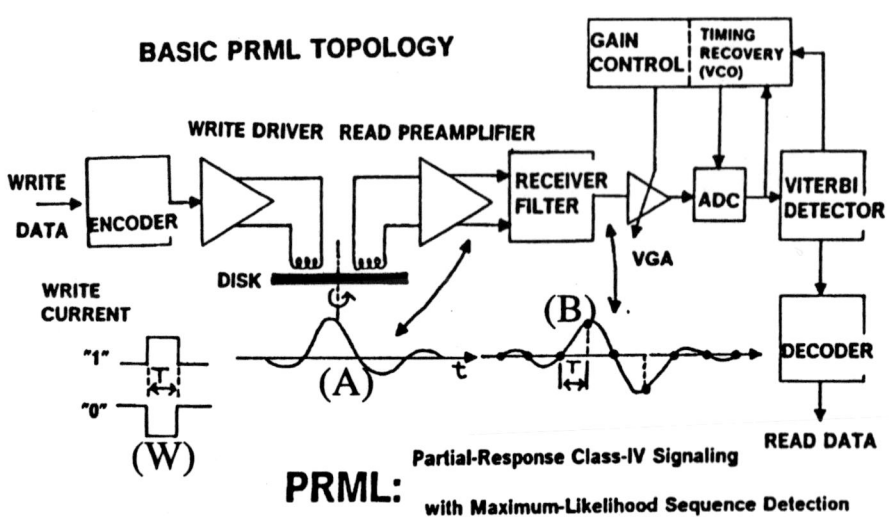

FIGURE 12.1 Product Design example architecture.

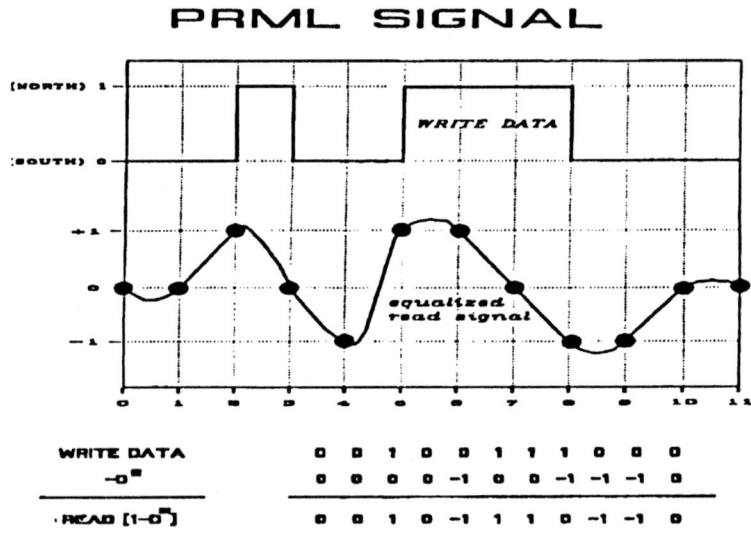

FIGURE 12.2 Product example overlap of successive waveforms.

Design of a Mixed-Signal IC

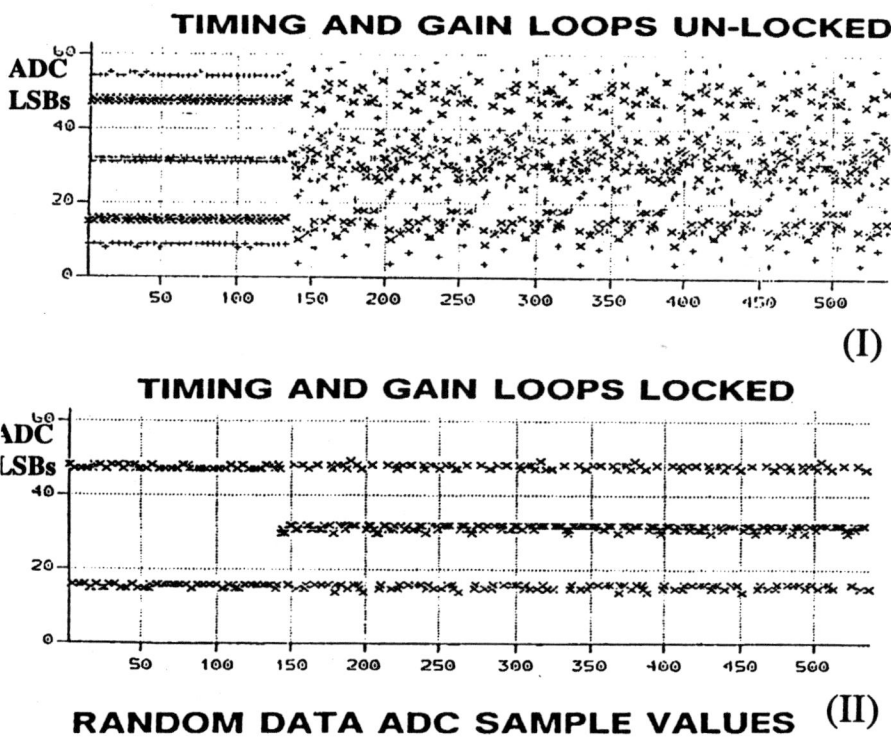

FIGURE 12.3 Action of Timing and Gain Loops.

Figure 12.3 waveform (I) shows what the sampled data looks like before the timing and gain loops are locked and sampling data at the proper amplitude and correct phase. Figure 12.3 waveform (II) shows the digitally sampled waveform at the proper amplitude and sampling phasing. The samples assume the three distinct sampling values of 1, 0, -1 as expected.

The IC contains all required analog and digital functions for a complete read channel, with the exception of a read head pre-amplifier to amplify the microvolt disk signal into the millivolt range. Write support functions include write precompensation circuitry, differential analog write driver interface to the inductive write head driver, and a programmable 4-bit write current bias source to the inductive write head driver. The

A Design Example

write pre-compensation circuitry allows the programmable shifting of selected data edges in 1ns steps to predistort data prior to write so that it has the proper timing relationships when read back. The IC block diagram is shown in Figure 12.4. Reference [12.11] details the semiconductor process and overviews the floorplan approach to mixing analog and digital on this IC.

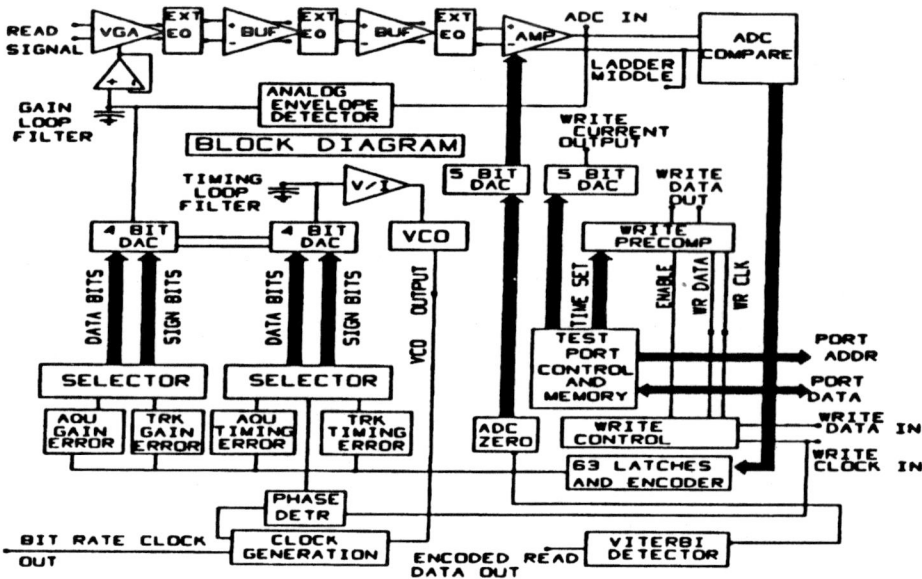

FIGURE 12.4 Product example block diagram.

One of the almost insurmountable problems encountered in this design was architecting the system to allow analog low level signals to coexist on the chip with noisy CMOS gates and drivers. Chip/package electrical modeling was key to obtaining a working system since CMOS switching noise couples to the analog circuits through the degeneratively doped P+ substrate. Additionally, the primary package resonance happened to be very close to the third harmonic of the 27Mhz clock. The chip was designed to allow a shut down of all analog circuits except the converters and VCO, to patch in quiet analog circuits from another IC. This was initially done to determine

Design of a Mixed-Signal IC

how much S/N (error rate) was being lost by merging the analog and digital circuits together on one IC. The comparison showed an almost insignificant 1/2 order error rate loss. This is very surprising considering that the chip logic and analog power busses showed roughly 1 volt and .4 volt noise, respectively, with significant frequency content above 100Mhz. Figure 12.5 shows how the various power supply noise levels changed with switching power in two different power modes. The READ mode had the highest power and noise and the WRITE mode had the lowest power and noise.

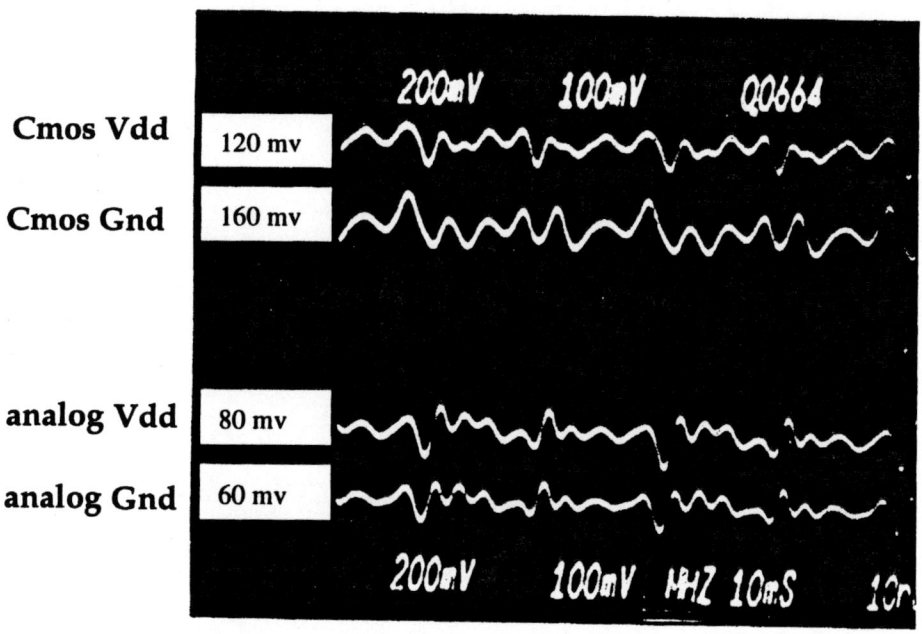

Switching activity	Switching Power	Cmos Vdd	Cmos Gnd	Analog Vdd	Analog Gnd
Max (read)	450 mwatt	650mvp-p	350mvp-p	350mvp-p	280mvp-p
Min (write)	100 mwatt	120mvp-p	160mvp-p	80mvp-p	60mvp-p

FIGURE 12.5 Minimum noise corresponded with minimum switching activity.

A Design Example   **259**

A Design Example

The chip is architected so that all loop filters are integrating in nature to cancel periodic noise. Figure 12.6 shows the architecture of the current output D/A converters that drive the loop filters. The timing and gain DACs are made from binary weighted combinations of current switches. A single current switch is shown in Figure 12.7. The power supply and substrate noise caused considerable periodic noise currents at the output of the D/A converters. These converter outputs provided the control signals that varied the gain in the AGC and VCO circuits. By providing a capacitor for the output loop filter, integration of the periodic noise was possible. The result was very little loop jitter in gain or timing.

Figure 12.8 shows the topology of the ADC converter used on this IC. The input of the ADC was a differential to single ended converter. This was to be able to reject common mode noise on the input. The converter also centered the ADC single ended signal in the middle of the resistor ladder to provide the capability for bidirectional signals.

Design of a Mixed-Signal IC

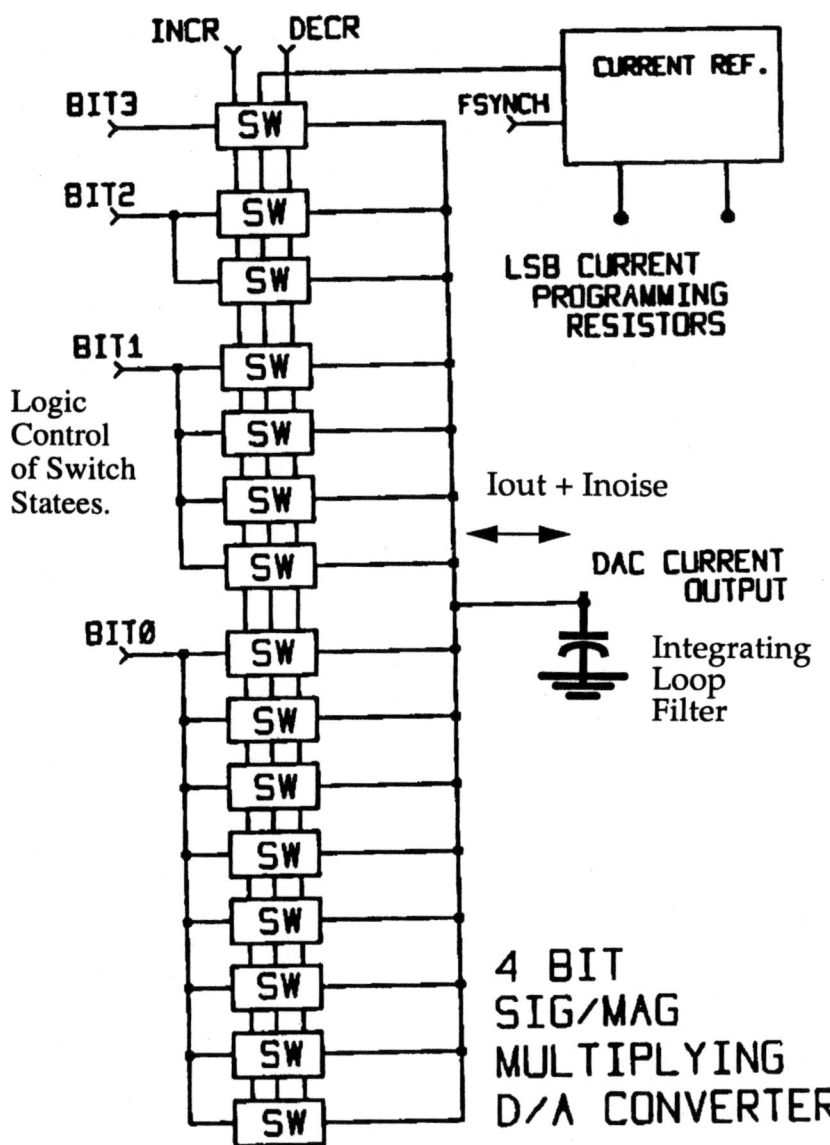

FIGURE 12.6 D/A Converter Block Diagram.

A Design Example

The rest of the ADC is a classic FLASH analog to digital converter. Figure 12.9 shows the waveforms at the top and bottom references of the resistor ladder as well as the center of the resistor ladder and the voltage after the converter at node A. Since the top and bottom reference and Node A are referenced to the VDD node of the IC they carry the power supply noise waveform of the power rail.

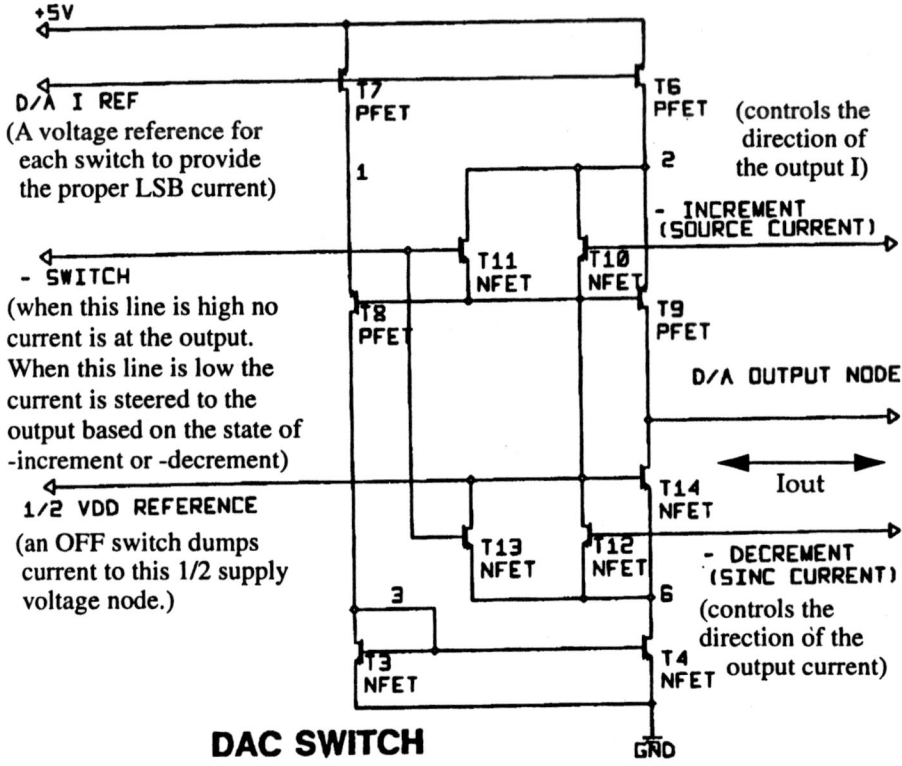

FIGURE 12.7 Dac switch.

Note that node A actually amplifies this waveform somewhat. However the voltage at the center of the resistor ladder sees the low-pass filtering effect of the resistor ladder with its associated capacitance at each resistor tap. The result is that the supposed common mode supply bounce is converted to a differential mode signal. In future designs this ADC was converted to a fully differential design throughout to cancel this noise effect. For this design the noise needed to be nulled out.

The chip read and ADC sample clock is derived from a VCO controlled by a timing loop. The VCO, shown in Figure 12.8, runs nominally at 54Mhz. The VCO frequency is divided by two, before use, to make the clock independent of up/down time waveform skew. A variable current to the VCO control voltage capacitor is provided from the timing loop via a 5-bit D/A converter, loop filter, and voltage to current translator sets the VCO oscillation frequency.

A Design Example

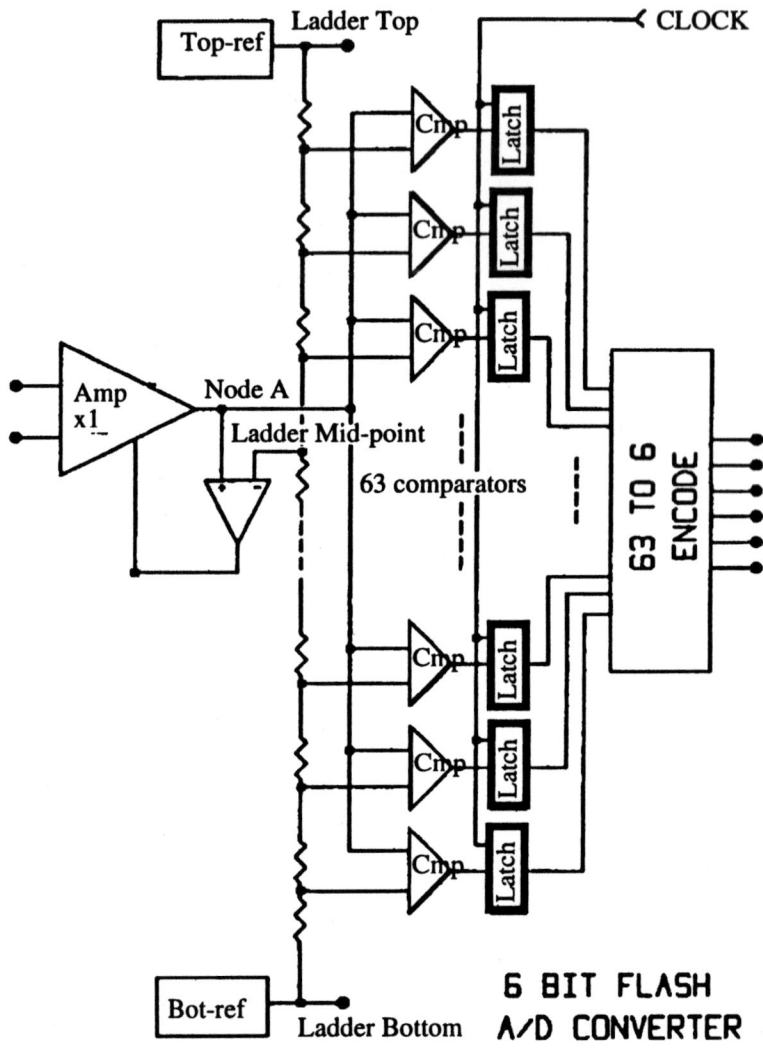

FIGURE 12.8 Problem ADC Topology.

In read mode the asynchronous write clock is degated at the chip receiver so it does not cause beat frequency problems or start the package resonating out of time phase

Design of a Mixed-Signal IC

with the ADC sample clock. As long as no asynchronous clocks are present, the package resonance noise reaches steady state. Now all the synchronous noise components manifest themselves as a fixed offset shift in the samples. The ADC auto zero circuit can then remove the offset and keep the total ADC error to <1/2 LSB. It has a +/- 4LSB cancellation range in 1/4 LSB steps. The value of the fixed offset will vary with power supply voltage, temperature, and switching activity, so the auto-zero is performed under use conditions at all disk idle times. Figure 12.10 shows that the resonant noise results in essentially a dc offset at sample times. Figure 12.11 shows the result of a beat frequency test on the ADC while being clocked at speed. The result is that noise voltages of up to 8 LSBs can be effectively nulled out to produce a converter with no missing codes.

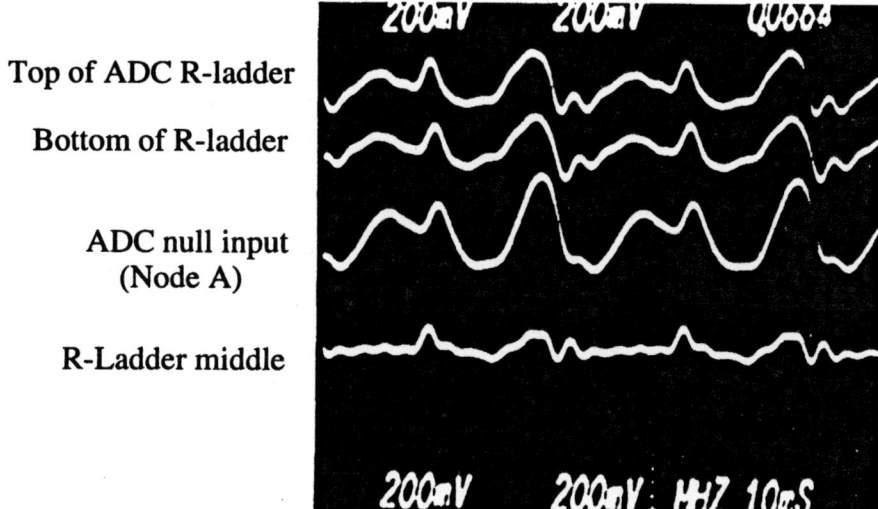

FIGURE 12.9 Common mode to differential mode conversion.

A Design Example 265

A Design Example

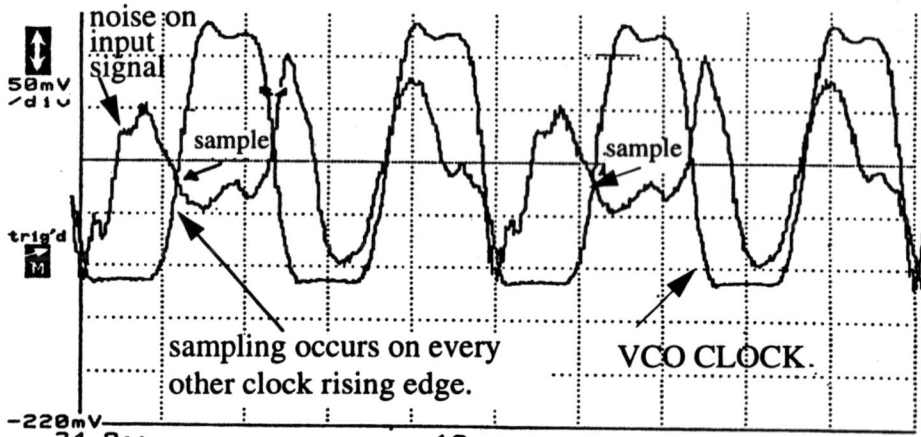

FIGURE 12.10 Relationship of Synchronous Noise.

ADC MISSING CODES

Module 3 A/D Samples from a 900 mVp–p, 6.765 MHz sinewave (Temp: 32 Celsius)

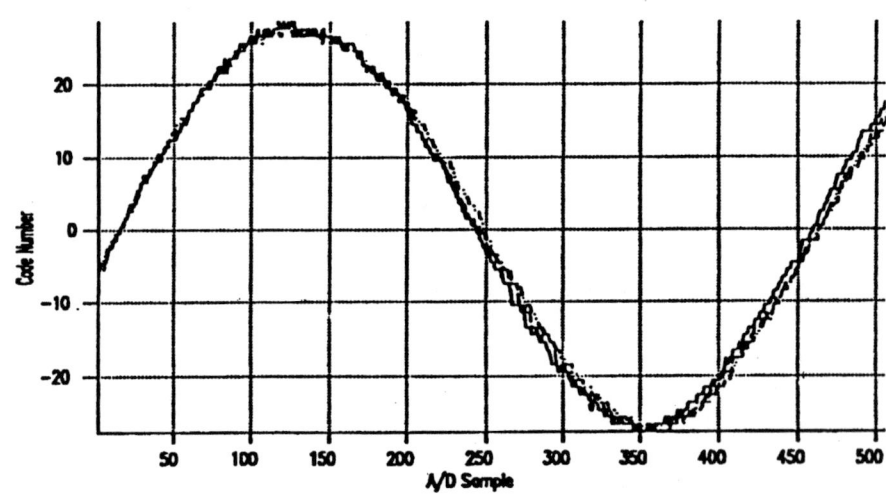

FIGURE 12.11 Net Affect of "Synchronous" Noise on ADC samples.

Design of a Mixed-Signal IC

The VGA (variable gain amplifier) accepts a read signal as low as 20mV with harmonics of interest to 13.5Mhz. The digital gain control loop drives a 5-bit current output D/A whose output ties to the gain loop filter capacitor. The output of the VGA drives the first stage of a three stage passive equalizer. Each stage is buffered by a fully differential voltage amplifier. The last section of the three section passive equalizer is a low pass filter at 13.5Mhz which removes all high frequency noise prior to the A/D pre-amp. Figure 12.12 shows the coupled noise on the AGC output before and after the anti-aliasing filter. By placing the package resonant frequency outside the Nyquist sampling frequency the noise could be reduced to 2% total harmonic distortion. Figure 12.13 shows how the tuning of the chip resonance was obtained for first prototype hardware before the capacitance could be included on chip. In write mode all read mode circuits (>90%) are shut down so they do not contribute to write data jitter.

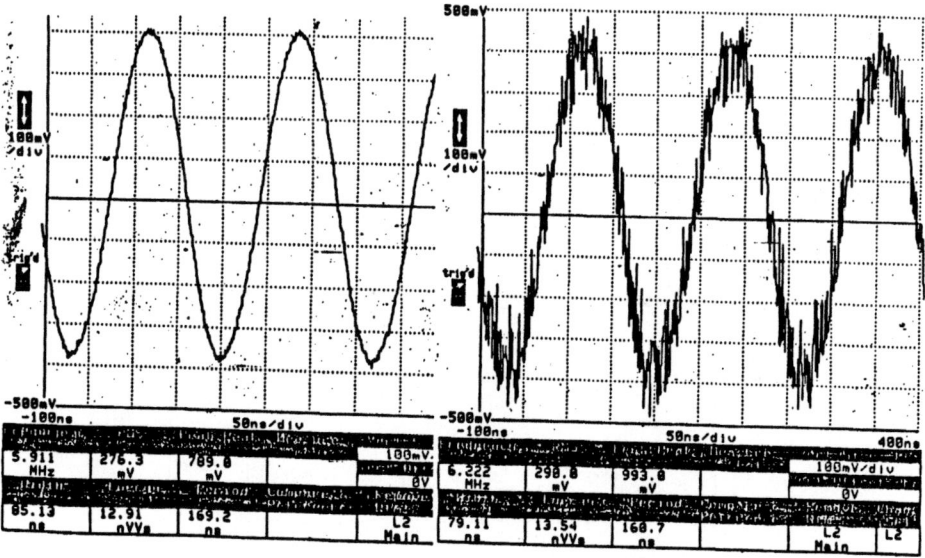

FIGURE 12.12 AGC output without (left) and with (right) logic switching.

Figure 8.12 shows the complete RLC chip and package model with values for this IC in a 68-pin PLCC. The chip bus wire inductance can be ignored due to its close proximity to the substrate (ground) and periodic decoupling capacitance (CLC) on chip. The N-Well diffusion in the P-Minus epi is responsible for CLV; N-Fet sources in the

A Design Example

P-Minus epi are responsible for CLG. Substrate contacts associated with each logic circuit form RLGS. At the center of the schematic diagram is the highly doped substrate node, SUB. The substrate underneath the entire chip is considered an equal-potential surface. The circuit model provides an independent path from the chip substrate to card-gnd. Every time the clock switches, logic circuits draw current from the logic power supplies and stimulate the power R-L-C circuit. The circuit rings out in a damped sinusoid for up to 100 ns, which is several times longer than the stimulation that came from the logic circuits. If the chip substrate is connected to logic-gnd through a resistor such as RLGS, the substrate will have all of the power bus bounce (noise) that is on logic-gnd. If the substrate contacts are removed from each logic circuit and there is no Ohmic path from logic-gnd to the substrate, the substrate will be coupled to the logic power supplies through just two paths: CLV and CLG. Logic-Vdd and logic-gnd ring 180 degrees out of phase if package impedances are symmetrical. By making CLV and CLG similar values, it is possible to couple equal amounts of out-of-phase noise to the chip substrate. Latchup is avoided by connecting a substrate ring at the edge of the chip (node SUBRING) to card-gnd through a package pin that does not carry power supply current. Chip substrate noise from the logic power supplies is minimized by this technique.

FIGURE 12.13 The Value of "ON-CHIP" Decoupling

Figure 12.14 shows the floor plan of a more complicated disk drive channel chip built on a P- substrate instead of a P+ substrate. The chip was split into four isolation areas or islands: (A) analog write functions, (B) analog read functions, (C) CMOS write functions, and (D) CMOS read functions. The off chip drivers on the periphery of the chip used the power domains of their adjacent islands except for the high speed, 1 volt swing, pseudo ecl drivers. They were put on their own power domain (E) but were not separated from the rest of the chip by an isolation region because of area constraints. In addition island (B) was broken into two power domains. A high speed A/D converter was placed in this power domain and its output section included some high speed CMOS logic that could not be easily separated from the analog front end of the ADC. The high speed logic section of the ADC was placed on its own power rails within island (B). The net result is that the chip had 6 distinct and separate power domains to model. Figure 12.15 shows a very simplified view of what the chip/package model looks like. Each island has substrate contacts distributed on the ground power bus. It is assumed that the composite resistance of the substrate contacts is small and effectively makes the local substrate under each island shorted to the local ground bus. A matrix of substrate isolation resistances from any island to any other island is theoretically required. This would require an island to island matrix resistive model with a resistor between every island to every other island. In the example given with a very large dominant island like (D) a simplification can be made with a resistor from (D) to every other island. Island (B) does not have substrate contacts on the ADC switching ground bus so as not to couple noise resistively to the chip substrate. The Island (B) quiet power bus for the ADC front end does have substrate contacts on the ground bus. Table 12.1 shows the resistance achieved by the island to island separation in this particular design. The separation was achieved by blocking all implants to obtain native 10 ohm-cm substrate in the separation regions. Metal 1 and poly lines were not allowed to cross this region. The design used four levels of metal interconnect.

A Design Example

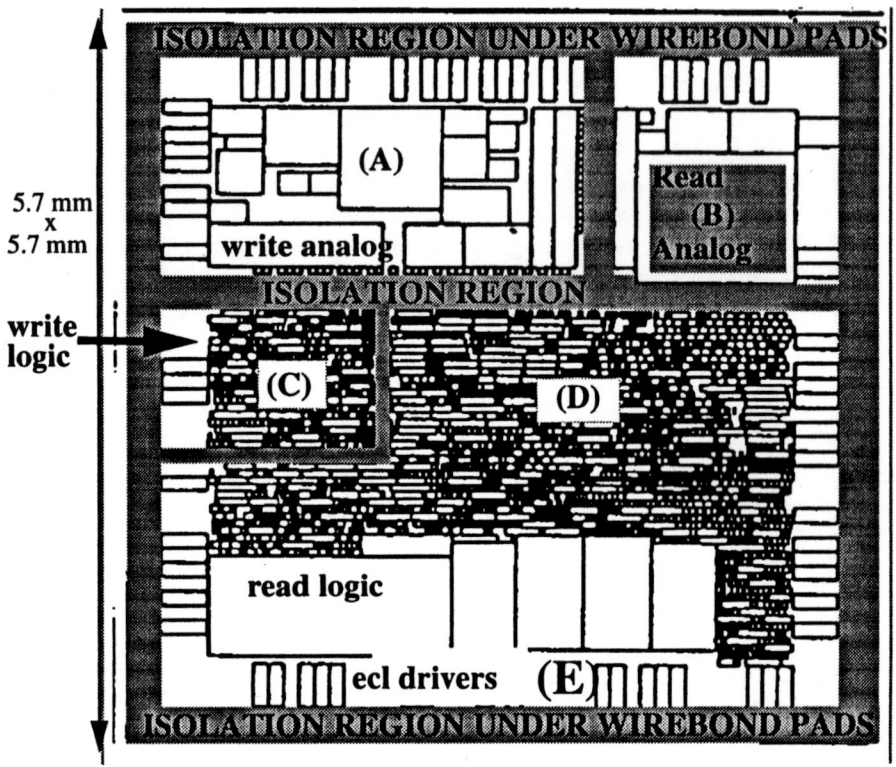

FIGURE 12.14 P- Chip Floorplan with isolation regions shaded.

270 Simulation Techniques and Solutions for Mixed-Signal Coupling in ICs

Design of a Mixed-Signal IC

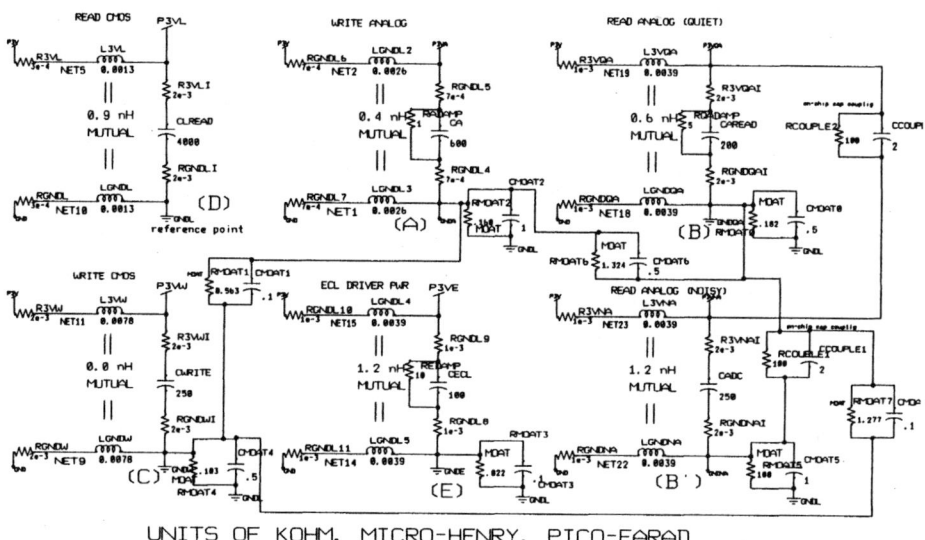

FIGURE 12.15 Simplified Equivalent Chip/Package model for Bus to Bus coupling.

TABLE 12.1 Resistance of island to island separation areas.

	Island A	Island B	Island C	Island D	Island E
Island A		160 ohm	176 ohm	160 ohm	180 ohm
Island B			233 ohm	182 ohm	202 ohm
Island C				103 ohm	124 ohm
Island D					22 ohm

12.1.1 Limitations to Resistive "moat" isolation

Since every signal line crossing the isolation region provides a capacitive connection to the power supply rails on each side, it provides a capacitance that is in parallel with the isolation resistance. Each capacitance has a series resistance which limits its shorting effect somewhat. If enough lines cross the isolation area then their parallel capacitances will increase to be quite large and their series resistances will parallel to

A Design Example 271

be quite small. The net effect is that the isolation resistance between circuit islands will be dramatically reduced in the frequency range where the shunting capacitive impedances are small. With this effect taken into account, the optimum isolation resistance between islands is often only a few tens of ohms. Resistances that are any higher have little benefit due to the shunting capacitive impedance and only take up more chip area. If a sensitive area of circuitry can be architected to have very few communicating signal lines to other circuit islands then the effect of the shunting capacitance can be dramatically reduced. Also, if many of the communicating lines are slow speed, they can be routed on polysilicon to increase the resistance in series with the parasitic capacitance to the power rails. Special buffers, designed to have very low capacitance from signal inputs and/or outputs to the supply rails can be designed and used. For CMOS buffers this can be accomplished by using the smallest MOS devices for the inputs and outputs of the buffers and by cascoding more devices in the CMOS inverters.

12.2 Summary

This chapter described a specific design example to which some of the previously recommended noise reduction techniques werw applied to optimize the design. The overall architecture and chip fllorplan were discussed and results were shown to indicate the extent of noise coupling on the IC and its effects.

REFERENCES

[12.9] Timothy Schmerbeck, "Mechanisms and Effects of Noise Coupling in Mixed Signal ICs," EPFL, Switzerland course presentation, June 29-July 10, 1992.

[12.10] T. Schmerbeck, et al., "A 27 Mhz Mixed-Signal Magnetic Recording Channel DSP Using PRML," *Technical Digest of IEEE ISSCC,* pp. 136-137, Feb. 1991.

[12.11] L. D. Smith, *et al.,* "A CMOS-Based Analog Standard Cell Product Family," *IEEE Journal of Solid-State Circuits*, Vol. 24, No. 2, pp. 370-379, April 1989.

Summary

[12.12] L. D. Smith, T. Schmerbeck, et al., "A CMOS-Based Analog Standard Cell," *IEEE Journal of Solid-State Circuits*, vol. 24, No. 2, pp. April 1989.

[12.13] R. Philpott, T. Schmerbeck, et al., "A 7Mbyte/sec (65Mhz), Mixed Signal Magnetic Recording Channel DSP Using Partial Response Signalling with Maximum Likelihood Detection," *IEEE Custom Integrated Circuits Conference*, May 1993.

[12.14] T. Schmerbeck, "Design Strategies for Reducing the Effects of Noise Coupling in Analog and Mixed-Mode ICs," in presentation for course on *Practical Aspects of Analog and Mixed-Mode IC Design*, Beaverton Oregon, May 18, 1993.

[12.15] F. Dolivo, et al., "Performance and Sensitivity Analysis of ML Sequence Detection on Magnetic Recording Channels," *IEEE Trans. Magn.*, Vol. 25, No. 5, pp. 4072-4074, Sept. 1989.

[12.16] T. Schmerbeck., "Minimizing Mixed-Signal Coupling and Interaction," *Technical Digest of IEEE ESSCIRC Conference*, Ulm, Germany, Sept. 1994, pp. 28-37.

APPENDIX A — Mesh Moments

The state equation for an RC circuit for step voltage excitations can be written as

$$C\dot{V}(t) = -GV(t) + B_1\dot{u}(t) + B_0 u(t) \tag{A.1}$$

where $V(t)$ is the vector of node voltages and $u(t)$ is the input step voltage excitation vector. C and G are the capacitance and conductance matrices. The matrices $B1$ and Bo represent the influence of the input excitation vector and its derivative vector on the node voltages in the circuit. In the frequency domain:

$$V(s) = (sC + G)^{-1}(B_0 + sB_1)u(s) \tag{A.2}$$

$$V(s) = [G(sG^{-1}C + I)]^{-1}(B_0 + sB_1)u(s). \tag{A.3}$$

Since $u(s) = s^{-1}$,

$$V(s) = \left[I - sG^{-1}C + s^2(G^{-1}C)^2 - s^3(G^{-1}C)^3 + \ldots\right] G^{-1}\left(B_0 s^{-1} + B\right). \tag{A.4}$$

The steady-state node voltage vector, $V(\infty)$ can be expressed as

$$V(\infty) = \lim_{s \to 0} sV(s) = G^{-1}B_0 \tag{A.5}$$

Mesh Moments

$$B_0 = GV_\infty. \tag{A.6}$$

Also the initial state node voltage vector, $V(0_+)$ is given by

$$\lim_{s \to \infty} sV(s) = C^{-1}B_1 \tag{A.7}$$

$$B_1 = CV(0_+). \tag{A.8}$$

Substituting (A.8) and (A.6) in (A.4) results in

$$V(s) = \left[I - sG^{-1}C + s^2\left(G^{-1}C\right)^2 - s^3\left(G^{-1}C\right)^3 + \ldots\right]G^{-1}\left(GV(\infty)s^{-1} + CV(0_+)\right) \tag{A.9}$$

or

$$V(s) = \frac{V(\infty)}{s} + [I - sG^{-1}C + \ldots]G^{-1}CV(0_+) - \left[G^{-1}C - \left(s\left(G^{-1}C\right)^2 + \ldots\right)\right]V(\infty). \tag{A.10}$$

Equation (A.10) can be rewritten as

$$V(s) = \frac{V(0_+)}{s} - \frac{(V(0_+) - V(\infty))}{s} + G^{-1}C(V(0_+) - V(\infty))$$
$$- \left(G^{-1}C\right)^2 s(V(0_+) - V(\infty)) + \left(G^{-1}C\right)^3 s^2(V(0_+) - V(\infty)) - \ldots \tag{A.11}$$

and comparing (A.11) to the moment equation for $V(s)$

$$V(s) = (D + m_{-1})s^{-1} + m_0 + m_1 s + m_2 s^2 + \ldots \tag{A.12}$$

where D represents any direct coupling between the input and output (i.e., outputs when all internal states are zero), gives

$$D = V(0_+) \tag{A.13}$$

$$m_{-1} = -(V(0_+) - V(\infty)). \tag{A.14}$$

Moreover all the higher order moments can be computed from m_{-1} using the recursion

$$Gm_i = -Cm_{i-1}. \tag{A.15}$$

APPENDIX B *Convergence Behaviour of Iterative Methods*

From the matrix equation

$$(D + L + U)x = b \tag{B.1}$$

where the matrix being solved for, A is decomposed as $A = D+L+U$ we obtained the Gauss-Jacobi iteration given by (4.95) as follows

$$Dx_{i+1} = -(L+U)x_i + b. \tag{B.2}$$

The error in the ith iteration is defined to be $e_i = x - x_i$. To determine how e_i evolves from one iteration to the next we subtract (B.2) from (B.1) to get

$$D(x - x_{i+1}) = -(L+U)(x - x_i) \tag{B.3}$$

which can also be written as

$$e_{i+1} = -D^{-1}(L+U)e_i = M_{GJ}e_i \tag{B.4}$$

where the matrix M_{GJ} is the iteration matrix for the Gauss-Jacobi method. Equation (B.4) can be written in recursive form as

$$e_i = M_{GJ}^i e_o. \tag{B.5}$$

We now switch from the ordinary Euclidean coordinate system to a new basis, which is given by the eigenvectors of the matrix M_{GJ}. Note that the eigenvector of a matrix, M is a vector, v which satisfies the following equation

$$Mv = \lambda v \tag{B.6}$$

where λ is an eigenvalue of M. Collecting the eigenvectors of M_{GJ} into columns of a matrix T and making the transformations,

$$e_0 = T\varepsilon_0 \tag{B.7}$$

$$e_i = T\varepsilon_i \tag{B.8}$$

we find that

$$\varepsilon_i = T^{-1}M_{GJ}^i T\varepsilon_0. \tag{B.9}$$

Since

$$M_{GJ}T = T\Lambda \tag{B.10}$$

where Λ is the diagonal matrix of the eigenvalues, λ_i of M_{GJ}, we find that

$$\Lambda = T^{-1}M_{GJ}T \tag{B.11}$$

and

$$\Lambda^q = \left(T^{-1}M_{GJ}T\right)\left(T^{-1}M_{GJ}T\right)\ldots\left(T^{-1}M_{GJ}T\right) = T^{-1}M_{GJ}^q T. \tag{B.12}$$

Substituting (B.12) in (B.9) gives

$$\varepsilon_i = \Lambda^i \varepsilon_0. \tag{B.13}$$

In the Gauss-Seidel iteration

$$(L+D)x_{i+1} = -Ux_i + b \tag{B.14}$$

it is easy to see that the iteration matrix, M_{GS} is given by

$$M_{GS} = -(L+D)^{-1}U. \tag{B.15}$$

Index

A
Admittance macromodel 85, 92, 100
Asymptotic Waveform Evaluation 87
AWE 87
AWE macromodel 92

B
BiCMOS process 161

C
Capacitive coupling 30
Chip, package shielding
 common mode radiation 238
 differential mode radiation 236
 far field emissions 235
Chip,package shielding 235
CMOS latchup 36
Cramer's rule 86

D
DC macromodel 100
Design example 255
Device models 47
 bandgap narrowing 49
 carrier mobility 48
 recombination and generation 51
Device noise 5–10

 avalanche noise 9
 burst noise 9
 excess thermal noise 7
 flicker noise 8
 shot noise 7
 thermal noise 5
Device simulation 43

F
Field penetration depth 189–192
Finite difference method 53, 80
Finite element method 53

G
Gummel's algorithm 63

I
Inductance 11
 flat wire 18
 mutual inductance 22–27
 round conductors 15
 self inductance 15–22
Inductive coupling 11

K
Kelvin ground 199, 225

M
Macromodel 85, 92, 100, 135, 141
Matrix solution 101
   gradient descent method 103
   ICCG 106
   iterative methods 101
   strongly implicit procedures 103
MEDICI 67, 107
Mesh generation 117
   a priori refinement 120
   adaptive refinement 117
   grid distribution 122
Micro-strip wires 31
Moat isolation 221, 269
   limitations 271
Muller's algorithm 86

N
Newton method 64
Noise coupling control 203, 226
   card layout 248
   ciruit topology 249
   effect of shielding 240
   ESD considerations 230
   noise cancellation 231
   packaging 244
   referencing 248

O
Off chip driver coupling 177

P
Packages 12, 13, 14, 29, 207–213
   flip chip 212–213
     metallized ceramic Pin-Grid-Array (MC-PGA) 207
     Metric Quad Flat Pack (MQFP) 14
     multi layer MC-PGA 208
     Plastic Dual-In-line (PDIP) 13
     Plastic Quad Flat Pack (PQFP) 14
   PLCC 12, 29
   split power feeds 155
   TAB 211
Power bus coupling 149
Power bus structures 150–155
   RANDOM 150
   simple grid 151–153
   STAR 150
   STAR using grids 154
   TREE 150, 153

R
Resonance 158

S
Scharfetter-Gummel 59
Skin depth 189–192
Substrate characteristics 188, 217
Substrate coupling 34
   coupling control 183, 203, 217
   effect of substrate bias 194
   ESD considerations 38
   inductive effect 36, 196
   latchup considerations 36
   modified single node model 129
   resistance extraction 135
   simple P- model 218
   simulation results 67, 107
   single node model 127, 160, 167, 176, 206–213
Substrate macromodel 92, 100
Substrate model 78, 92, 100, 127, 129, 160, 167, 176, 206–213, 218
Substrate resistance extraction 135
   interpolated macromodel 141
   nested macromodel 135
Substrate splitting 227, 228
Switching logic model 166
Switching noise 10

W
Well isolation 229, 242